Data: Mirrors of Science

Data: Mirrors of Science

by
R. HOUWINK
Wassenaar, The Netherlands

AMERICAN ELSEVIER PUBLISHING COMPANY, Inc.
New York 1970

AMERICAN ELSEVIER PUBLISHING COMPANY, INC.
52 VANDERBILT AVENUE
NEW YORK, N.Y. 10017

ELSEVIER PUBLISHING COMPANY, LTD.
BARKING, ESSEX, ENGLAND

ELSEVIER PUBLISHING COMPANY
335 JAN VAN GALENSTRAAT, P.O. BOX 211
AMSTERDAM, THE NETHERLANDS

Standard Book Number 444-00068-2

Library of Congress Card Number 76-100399

PRINTED IN THE UNITED STATES OF AMERICA

Why This Book?

It requires a word of introduction to bring Dr. Cipher before the readers of this book. He is, in fact, an extreme type of the dry pedagogue: "Young Johnny" is one of his pupils.*

Thus the Doctor: "For me, numbers bring light to the darkest problem; but as for young Johnny, they leave him in obscurity."

This book—*Data: Mirrors of Science*—can be seen as a continuous confrontation of Dr. Cipher, that man of chilly science for whom numbers are adequate as the total expression of an idea, by young Johnny, his pupil, in whom may be recognized the compiler of this book. It will be seen that young Johnny is not always happy with "just numbers" and searches for the deeper meanings hidden behind them. For him numbers are often enough of a screen for the Dr. Ciphers of this world, those people who fail to consider the relationships of numbers to many other values which we encounter every day.

We are usually able, be it with the aid of instruments, to set bounds to Nature's proliferation and to determine her chosen proportions, but . . . what do such computations really mean to us? For example, what impression can we form of the size of water molecules when we know that a thimble can contain billions upon billions of them? And what does the astronomers' unit, the light-year of 10^{13} km (6×10^{12} miles) tell us?

To condense data, and thus to present them in a readily comprehensible and, especially, readily comparable form, we often find tables very useful. For some data, graphs may be an improvement on tables, but when proportionally large differences—more than a very few orders of magnitude (or, powers of 10)—are to be illustrated, we need something like a logarithmic scale, which, for many readers may destroy their impression of the relative proportions of the numbers under consideration.

* These persons are introduced to us in a well-known Dutch book by Frederik van Eeden (1860–1932), *The Little John.*

A more effective aid to the understanding of the scale of some of Nature's unimaginably small—and large—dimensions is the suggestion of familiar, or even unfamiliar, *images* that will convey an element of reality to the reader's mind and so assist understanding and comparison of the facts and figures.

If we learn that the daily intake of energy for an average man, 2000 kilocalories, is just about equivalent to that of a 100-watt electric lamp shining through a day as well as a whole night, we can forget how *much* energy they consume; it is the *comparison* that is significant. Images such as these help the reader to see clearly, and thus to remember, mutual relationships of—at first sight—unrelated quantities.

And if a scientist speaks of a "nanosecond"—one thousand-millionth of a second, or 10^{-9} second—does it not help us to realize his meaning if we can recall that as many nanoseconds pass in one second, as seconds have passed in the last 30 years?

So, let images be our allies in our endeavors to understand Nature better. It is surely no blemish that the choice of example in this book often enters the realm of the comic.

Wassenaar, The Netherlands, 1969 R. HOUWINK

Compiler's Note

This book is in some degree a continuation of *The Odd Book of Data*, published by the firm of Elsevier in 1965. The great interest aroused by that collection induced the compiler to extend his studies, and such a vast amount of fresh material was discovered that a new book came into existence. Although some entries from the former *Odd Book* may be found in this new volume, this is only because they are indispensable. It is, in fact, like a new history book with some old maps.

It is the compiler's wish that readers should not attempt to go right through the whole book at one sitting. Some entries are highly concentrated, and some include quite intricate reasoning. Some are, however, otherwise, so it is hoped that the scope of the book is wide enough for many to find something of interest in it.

Acknowledgments

A special word of thanks is due to

Mr. J. P. van Lier

Mr. M. Vertregt

both of The Hague, Holland, for their thorough and extensive advice concerning the scientific aspects underlying most of the entries.

Moreover I am much indebted to

Dr. E. F. J. Janetzky, of The Hague,

and

Mr. J. Janssen, of Delft

for their more special advice which has contributed much to the realization of the book.

I should like to express sincere thanks to

Mr. J. Jackson, of Condover, near Shrewsbury, England who has undertaken the necessary translation. Actually he did much more than a literal translation by revising many entries from a scientific viewpoint and offering some very useful new contributions.

In conclusion the author is highly indebted to Mr. E. J. Dekker, Voorburg, The Netherlands, for designing the illustrations with much imagination and skill.

Should this book be recognized as a valuable aid to the understanding of science, then a great deal of the credit should go to the persons named above.

R. HOUWINK

Contents

Chapter 1

Data and Images

Nothing can be more fatal to progress than a too confident reliance on mathematical symbols and data; for the student is only apt to take the easier course, and consider the data, and not the fact as the physical reality.
(Freely quoted after Lord Kelvin)

1.1 Converting Data into Images

It is perhaps interesting to devise graphic or pictorial comparisons and interpretations that will make numerical data more understandable and meaningful than just numbers alone.

As an example, we read about the distances of stars in terms of light-years—even the nearest is about $4\frac{1}{2}$ light-years away from the solar system—and, perhaps on the same page, about the 100 kilometers (km) or 60 miles, separating two islands.

Now the light-year, 10 million million (10^{13}) kilometers (km), is 10^{11} times the distance between the islands, and while most of us can recognize the mathematical convenience of expressing the very large ratio in this way, we suffer meanwhile from a sort of "imagination vacuum" on making any attempt to visualize what 10^{11} is, taken purely as a number and neglecting its deeper implications. However, an image which perhaps helps in this case is comparison of the thickness of a human hair with the diameter of the Earth, for which the ratio is also 10^{11}.

What is required of such images in order to make a ready appeal to the mind, may appear from the following instance concerning population densities.

Table A gives this information in the usual form, inhabitants per unit of area:

Table A
Population densities expressed in persons per square kilometer (km^2)

Region	*Persons per* km^2
World as a whole	25
United States	25
U.S.S.R.	10
United Kingdom	240
West Germany	240
New York City	9000

In the first four cases, this sort of presentation is adequate, for it is easy to visualize something like 10, or 25, or even 250 persons occupying a square kilometer—for the last, think of cows grazing in a large meadow or look at Fig. A. When we come to the cities, however, it is

Fig. A. One square kilometer occupied by 10, 25, or even 250 persons.

not easy to build up in our minds a picture of 9000 people on a square kilometer of ground, even starting from the crowd at a Cup Tie, or a "World Series."

Let us therefore express the same basic information in another way, namely by taking as the "yardstick" the average distance between individuals, when they are distributed in a uniform "square" pattern (at mutual distance d) over the square kilometer, as if they occupied the corners of squares on a chessboard (see Fig. B).

Fig. B. The average distance, D, between individuals.

Table B
Population Densities

Region	Distance to our nearest neighbors *In meters*	*In yards*
World as a whole	200	220
United States	200	220
U.S.S.R.	320	350
United Kingdom	70	77
West Germany	70	77
New York City	11	12

Everybody can "picture" in his mind a distance of a few hundred meters, with little more difficulty than "seeing" the New York distance of 11 meters. We may also note that whereas in Table A the "cities" situation was the most obscure, it becomes in Table B the easiest to visualize.

So the presentation in Table B is a considerable improvement over that in Table A for we have gained clarity in one part of the table without losing it elsewhere. Moreover the numbers in Table B and the whole image give us a more real and so more easily memorized picture.

We can easily imagine ourselves standing at a street intersection, seeing one of the Earth's population about 200 meters away along each route, or seeing our fellow New Yorkers about 10 meters away in each direction, hardly more than across the street.

Through this example we are now familiar with the idea of the second "power" of 1 kilometer, the meadow of 1 km^2 (0.4 square mile), and perhaps we feel that changing from the second power to the third—from 1 km^2 to 1 km^3 (0.25 cubic mile)—will be an easy step for our imaginative powers. Alas, most of us have no notion whatsoever of the real scale of 1 km^3.

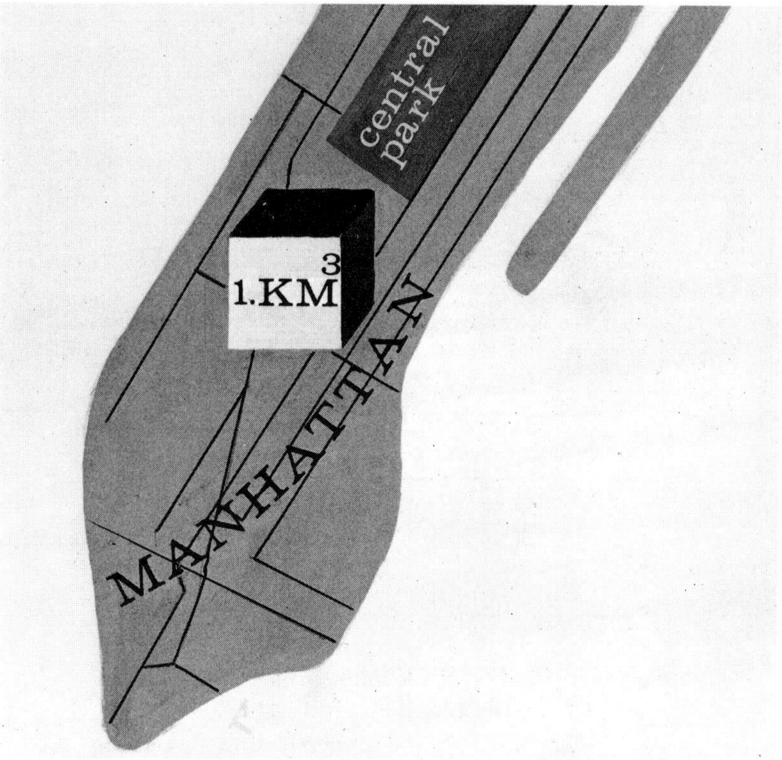

. . . the World's whole population in one-quarter of 1 cubic kilometer (0.06 cubic mile) . . .

How many have appreciated that if the world's merchant navies, including oil and other tankers, were to discharge their combined cargoes into a space of 1 km^3, this space would be less than $\frac{1}{3}$ full at the end of the operation? Or that, if we assembled the world's whole population in the same cubic kilometer, mysterious because so difficult to visualize, hardly one-quarter of its volume would be filled?

So we illustrate the dangers of thinking in terms of nothing but numbers and neglecting their deeper implications.

1.2 Easy to Keep in Mind . . .

● Keep in mind the number 3.5×10^9 (as an average between 3000 and 4000 million) and you have an idea of the order of magnitude of:
- the number of years which have elapsed since the earth solidified
- the number of people on the earth
- the number of cells in a man's brain
- the number of heart beats in a man's life.

● Keep in mind the numbers ranging between 10^{11} and 10^{12} and you will have an idea of:
- the ratio between the diameter of the Earth and that of a human hair
- the ratio between the weight of the world's largest liner ever built, the Queen Elizabeth (83,000 brt) and that of a bee (1 gram)
- the number of gallons in a cubic mile (9×10^{11})

● Keep in mind the numbers ranging between 10^{12} and 10^{13} and you will have an idea of:
- the number of cubic kilometers, contained in the Earth (10^{12})
- the annual electricity consumption in the world expressed in kilowatt-hours (4×10^{12})
- the number of kilometers in a light-year (10^{13})

● Keep in mind the number 10^{23} and you have an idea of the order of magnitude of:
- the number of molecules in a cubic centimeter of water
- the number of red blood corpuscles on earth
- the number of stars and other heavenly bodies in the observable universe (actually this is $\frac{1}{10}$ of the number mentioned).

● The average annual rainfall over the whole surface of the Earth happens to be equal to the height of the mercury column in a common barometer (760 mm).

● The speed of light is a million times that of sound. The speed of sound is a million times that of smell (see Section 7.1).

● New York City covers as many square miles as there are days in a year (365.4 square miles).

Chapter 2

Mathematics and Fundamental Physics

The truth has never been of any real value—it is a symbol for mathematicians and philosophers to pursue.
(Graham Greene)

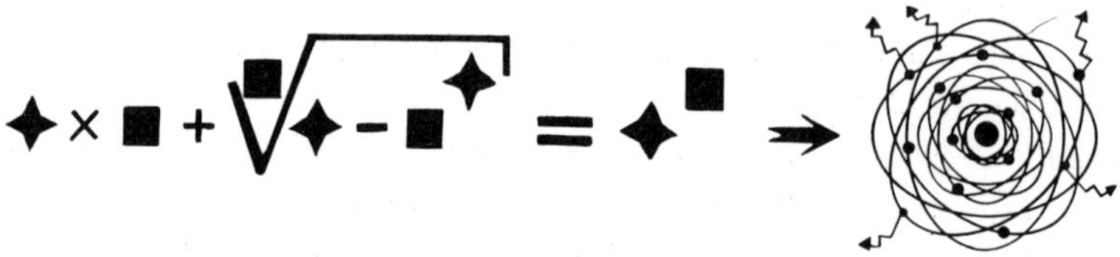

2.1 Giga, Tera—Nano, Pico

Gigawatts and terawatts, nanoseconds and picoseconds—these are units which, with ever-widening scientific developments, are coming more and more into use.

What they really signify can be illustrated thus:

giga $= 10^9$
nano $= 10^{-9}$ There are as many watts in a "gigawatt," and as many "nanoseconds" in 1 second, as the number of seconds since the time of Hitler's final preparations for the Second World War, thus in a period of 30 years.

tera $= 10^{12}$
pico $= 10^{-12}$ There are as many watts in a "terawatt," and as many "picoseconds" in one second, as ten times the number of seconds since the building of the Pyramids in Egypt, thus in 30,000 years.

By the way: One nanosecond is the time during which light travels 30 cm, and one picosecond the time during which light travels 0.3 mm.

2.2 Infinity—Far Away, Yet Still Close By [2]

The word "infinity" brings to our minds the measureless distances of the universe and the immense number of inaccessible heavenly bodies. But this "immense number" is not in fact uncountable, for it has been calculated that the universe as we know it comprises "only" 10^{80} separate particles.

However, in a liter of the air which we breathe, or of a gas in the general sense, the number of possible micro-states* is of the order of $10^{10^{20}}$ (or $10^{100 \text{ million million million}}$), which is a number far exceeding the 10^{80} components of the universe.

Suppose that we attempt to represent this number of micro-states by penciling dots on a piece of paper just far enough apart—say 1 mm—to be seen separately; how long will our piece of paper have to be? A whole light-year (9.5×10^{12} km) will suffice for "only" 10^{19} of our dots, so our record of micro-states will have to be as long as

$$\frac{10^{100 \text{ million million million}}}{10^{19}} \text{ light-years}$$

This may also be written as $10^{100 \text{ million million million}-19}$. But if we substract 19 from 100 million million million we are left with a number still for practical purposes 100 million million million, thus itself a fair representation of "infinity."

So, in 1 liter of air the concept "infinity" is realized in the number of possible micro-states which could exist in it. We have this close to us and around us at all times, but it remains a difficult conception.

In somewhat more familiar terms, and bringing visible reality a little nearer, we may recall that in a molecule of ammonia the molecular vibration is at the rate of 24×10^{9} per second; this scale of activity can be represented by every inhabitant of the Earth moving 8 fingers at the same instant.

2.3 Thinking with Einstein [3]

According to Einstein's relativity theory, *length*, *mass*, and *time* are dependent on the velocity of motion, relative to the observer.

Length

An observer "sees" the length of a moving body shorter in the direction of motion, the contraction being a function of its speed (without any stresses being involved):

* Micro-state defines a situation which is reproduced only when *all* the molecules in the system in question are in exactly the same positions relative to each other, as determined by location, direction of movement, and vibration state.

- in the case of an automobile (100 km per hour), the observer would not yet be conscious of the contraction, which corresponds to no more than the diameter of an atomic nucleus;
- the same is true of a jet plane (1000 km per hour), whose length should appear one atom smaller than it actually is;
- the apparent contraction of an interplanetary rocket traveling at 400,000 km per hour would amount to 0.01 mm;
- to an observer on another planet in the solar system the Earth would appear 6 cm shorter in the east–west direction, due to its rotational speed of 30 km per second about the sun;
- if the Earth were to rotate 10,000 times faster about the sun, a 1.8-meter (6 foot) man, lying in a bed placed parallel with the equator, would appear to the observer to be only 1 meter (3 feet) tall. If his bed were turned at right angles to the equator, he would seem to regain his original length, but the breadth of his body would then appear to be halved.

Mass

Mass increases with speed, becoming infinite at the velocity of light. If a medium-size passenger car with a stationary weight (mass) of 1000 kg (2200 lb) is accelerated from rest to a speed of 60 km (40 miles) per hour, it gains a fraction of a fraction of the mass of a pinhead (1 to 2 millionths of a gram) in the process.

If the car could be made to travel at a little over 100,000 km per hour—85 times the velocity of sound—it would indeed gain the mass of a pinhead (3 mg), if some means could be found to prevent its escape from the earth (escape velocity 40,000 km per hour).

At 250,000 km per hour the gain of 26 grams would be equivalent to that of a small bolt of the car. In order to gain its own mass, the speed should be about 2.9×10^5 km per second, 0.98 times the velocity of light. At extremely high speed (say 10^{-14}% below the velocity of light), a revolver bullet would have the mass of a fully laden freight car.

Time

To an outside observer, the passage of time is reduced by velocity. The return journey to a star, 9 light-years distant, by a rocket traveling with a speed of 10^{-8}% below the velocity of light, would take slightly more than 18 years. The heart and lungs of the rocket's crew, and the clock they carried with them, would appear, however, to the crew to work so slowly, that the 18 years would seem to last but a few hours. The observer would expect the rocket to have completed its two-way journey between, say, lunch and dinner on the same day, but at the same time he would observe that the families of the rocket's crew would have consumed lunch and dinner well over 6500 times.

Following again Einstein's theory, space is curved (non-Euclidean) in the vicinity of massive bodies. In such space the ratio between the circumference and the radius of a circle deviates from 2π.

If, after accurately measuring the circumference and radius of a large circle drawn on an open space, like Piccadilly Circus or Times Square, and thence determining π experimentally, we observe that a small insect has crawled into the circle, the mass of this creature would impart sufficient curvature to the surrounding space to alter, say, the thirty-fifth decimal place of π. The 765 remaining decimals, as calculated by a Danish schoolmaster (Section 9.2) would then of course become meaningless.

2.4 Uncertainties for Electrons and for Life [4]

The Heisenberg principle states that the momentum and position of a particle cannot simultaneously be exactly known.

If the uncertainty in position in any direction x is dx and the uncertainty in momentum in that direction is dp, then $dx.dp \geqslant h$, where h is Planck's constant.

For an electron this principle is of paramount importance. The single electron in a hydrogen atom in the ground state has a kinetic energy of motion about the nucleus of 13.6 eV, corresponding to a velocity of 2000 km per second. Its momentum is only 2×10^{-19} g cm per second and its position cannot be defined to better than 3×10^{-8} cm, a figure close to the size of the hydrogen atom. In fact one can fairly say that the uncertainty principle defines the effective size of atoms.

For ordinary life on the contrary Heisenberg's principle is not of much importance. If we know the velocity of a 1-ton lorry at a particular instance to 1 cm per second (about 0.1% at 35 km/hour) it would not matter that it was theoretically impossible to determine its position at the same instant more closely than 6.5×10^{-33} cm.
(Planck's constant $= 6.55 \times 10^{-27}$ erg sec)

2.5 Mobile Adsorption

"Mobile adsorption" describes a state in which (gas) molecules adsorbed on an active surface have the facility, or opportunity, to move over the surface of the adsorbent. In more detail, the energy of the adsorbed molecules must exceed the energy-potentials of the boundary atoms, while remaining less than the (overall) desorption energy.

When the inert gas xenon is adsorbed on a mercury surface, the xenon molecules appear to move about on the mercury and to be oscillating perpendicular to the surface. Again in more detail, this is seen as a sliding and hopping movement, with jumps of about 5Å—a distance slightly more than the distance between the mercury atoms in the surface—and of frequency 4×10^{11} per second.

This frequency is about the same as that of the electromagnetic waves which we call "infrared," but we are still a long way from producing a "xenon-mercury" radiation.

2.6 Frequencies

When we recall that the frequency—"γ"—of visible light is 10^{15} per second, we are saying that the number of vibrations in every second is "only" the number of seconds which have elapsed in the last 30 million years. If we equate the number of seconds which have elapsed since the formation of the Earth as a gas cloud, according to certain theories, 5000 million years ago, with a frequency (per second), then we find ourselves in the borderland between ultraviolet radiation and X-rays, where $\gamma = 1.5 \times 10^{17}$.

However, for "really" high-speed vibrations, such as we find in γ-rays with their frequency of 10^{22} per second, the age of the Earth would have to be 100,000 times the present estimate for its duration—in seconds—to equal the number of vibrations of γ-rays in *one* second.

Radio waves present quite a different picture at the other end of the spectrum, for, with a frequency as low as 100 per second for "long" waves, the interval comprising the same number of seconds is just under 2 minutes.

2.7 Gliding in Crystals [76]

When crystals are being deformed, lattice planes glide along each other due to the motion of linear lattice faults (dislocations) at tensions which are 100 to 1000 times smaller than those which would be necessary to deform a faultless crystal.

The length of the dislocation lines in an undeformed crystal can be between 10^6 and 10^7 cm per cm^3 with a distance between the atoms (ions) of say 2.5Å. Assuming that these dislocations are evenly distributed and that they all move in the same direction, one can obtain an impression of their number in the following way.

Enlarging a cubic centimeter of such a crystal to the size of a schoolroom ($10 \times 10 \times 10$ meters), the student would see the dislocation lines 0.5 cm apart, whereas to make the atoms "on" these lines become visible to him (i.e., have a diameter of 0.1 mm) he would have to use a magnifying glass producing a $400 \times$ enlargement.

2.8 Mass without Mass?

Some authorities assign to neutrinos a mass 1/1000 that of an electron, thus 9.1×10^{-28} gram, while others hold that neutrinos have no mass at all; all agree that neutrinos do not carry any electrical charge. However, neutrinos do possess energy in the forms of translational momentum and rotational momentum.

As to "translational" momentum, the sun is believed to emit a supertanker's load—100,000 tons—of neutrinos per second, of which only 50 grams reaches the Earth; but even this small fraction of the sun's output is an enormous number of neutrinos, being about 3 per square centimeter per second for each of us.

Yet, of this enormous number, it is thought that only a few score neutrinos are actually intercepted by the material of the Earth. The rest emerge "on the far side" of the Earth and pursue their way into space.

Considering that the absorption of other neutral particles—fast neutrons—requires something as thick as a few hundred volumes of *Encyclopaedia Britannica* (see Section 8.14), for neutrinos the Earth is exceptionally easy to pass through.

2.9 Forces of Attraction in the Universe [6]

Known to us now in Nature are four "interactions" which, while determining the ultimate form of the universe, differ to an extraordinary degree in their relative strengths, thus:

- *Gravitation*

This is the cause of what we call "weight," and while perhaps the earliest known, it is still the least well explained of the four interactions. On our Earth, we name the force of attraction on 1 cm^3 of water (at the Earth's surface) "1 gram."

- *The "Weak" Interaction*

This is the underlying cause of radioactivity. The postulated, but not yet found carrier particles have been named "W-bosons." To give an idea of the force which these particles exert on one another, imagine a cone cut out of the Earth, with its apex at the Earth's center and its base equal to the area of Brazil. If gravitation were as strong as "weak interaction," 1 cm^3 of water would have the weight which we now estimate for the "Brazilian cone." that is 10^{25} grams or 10^{19} tons.

- *Electromagnetic Interaction*

This originates from the arrangement of electrons in atoms, and is the underlying cause of "chemical" reactivity with, as carrier particles, photons. Such interactions are known to us as the van der Waals and the London forces. To give an idea of this force, we may say that if gravitation were as strong, then 1 cm^3 of water would have a weight 1600 million times that of the Earth (by itself 6×10^{21} tons, measured at the surface of a sphere of equal mass), thus 10^{31} tons.

- *The "Strong" Interaction*

This exists within the atomic nucleus, with the "pion" as carrier particle; it underlies "nuclear energy." If gravitation were as strong as this interaction, our cubic centimeter of water would weigh 100 times as much as the previous one, 10^{33} tons.

À Propos

If 2–3 grams of protons (the weight of a dime), were placed on each of the opposite poles of the Earth they would repel each other with a force of about 30 tons. There must thus exist within the atoms very great forces of attraction to hold nuclei together since they would otherwise explode.

2.10 The Power of Antimatter [7]

Early in the twentieth century, the scientist P. A. M. Dirac suggested the possible existence of "antimatter," composed of relatively heavy negatively charged "antiprotons," and of much lighter positively-charged "antielectrons" (or, "positrons" as these were later named), the weight ratio of these being about 2000 to 1 as is the case for ordinary protons and electrons.

In 1965, by the bombardment of the element beryllium with fast-moving protons, experimenters succeeded in forming "heavy" "anti-hydrogen" nuclei, composed of an antiproton and an antineutron. When these "antinuclei" were brought into contact with material composed of normal atoms, a violent liberation of energy resulted, apparently because of the transformation of "matter+anti-matter" into "energy."

It now seems likely that the meteorite which fell in a remote part of Siberia in 1908, terrifying the inhabitants over a wide area with a shattering explosion and a blinding flash of light, was in fact composed of this "antimatter."

As to the energy involved, an estimate based on the observed effects of the event, which included formation of a "crater" 7 to 10 km (4.5 to 6 miles) across, suggests that this energy was about the same as the energy that would be released by the explosion of 50 "Hiroshima" atomic bombs, in total 10^{23} ergs (see Section 12.6).

2.11 Short-Lived Particles [8]

In the strange and elusive world of the highly unstable elementary particles, the lifetimes under study are getting shorter and shorter. Such particles are usually moving at a speed approaching that of light (3×10^5 km per second).

One of the largest machines for producing high-velocity particles is in Russia, the 70 GeV* "Ring-tunnel" at Serpukhov with its circular path 1.5 km long. It takes a man 18 minutes, or 1100 seconds, to walk through it, once around. But when it is in action, protons are accelerated to a speed with takes them around 200,000 times in 1 second. So, in the time that a man has covered 1.5 km, the protons have covered a distance equal to 8000 times round the Earth.

* 1 GeV (giga electron volt) = 1000 million electron volts.

Particles of lifetime 10^{-10} sec, discovered in 1960, are now classed—based on the length of their track in the Wilson cloud chamber of 10 cm—as "slow" particles.

But those with a lifetime of only 10^{-33} sec, the so-called resonances, are now classed as "fast" particles: their traces in the cloud chamber are no longer than the width of an atomic nucleus.

So we may note that the proportion between the two shortest lifetimes for such particles, discovered now and in 1960, is as 1 second is to 100,000 times the age of the earth (since it solidified 3.5×10^9 years ago).

2.12 The Heaviest Material

If it were possible to constitute material from only the nuclei of atoms, thus stripped of their electrons, the material would have a specific gravity about 40,000, which is about 3000 times the specific gravity of lead. This may seem mere fantasy, but it appears that the well-known star Sirius has a "companion star" of which the specific gravity approximates . . . 40,000.

A quantity of electron-free nuclei whose total weight equaled that of the Earth (6×10^{21} tons) would have a volume of about 40% of the Atlantic Ocean (1.4×10^8 km^3 of nuclei).

2.13 A Captive Moon? [4]

If one could persuade the exclusively "nuclear" material, referred to in the previous entry, to distribute itself in threads, instead of only in "droplets," the strength of nuclear forces is such that a thread only 1 mm in diameter (with an adequate winch on the end) would be strong enough to pull the moon toward the earth, into a "24-hour," or synchronous, orbit, and hold it there.

With the moon fixed relative to the earth, one-half of the earth could dispense with most of its street and similar lighting, but the now permanently "dark" half would surely protest most strongly.

Another possible use of our "nuclear thread" would be to raise the land area of England and Wales—and 25 km of ground below it—out of the sea, thus perhaps solving the problem of "Channel Tunnel" versus "Channel Bridge," for neither would be needed.

Still more striking, the strength of intranuclear forces is so great that if a bag, say 2 meters in diameter, was formed from a "nuclear sheet" only 10^{-12} cm thick, a hydrogen bomb could be detonated within this bag without bursting it.

2.14 Speeds: from Cars to Gammas

An air of expectancy hung over Trafalgar Square. A huge crowd sized up eight competitors for a novel international event: "The One-Second Sprint."

The first participant, Mr. Car, the new national champion, flashed into the Square and steered four squealing tires through the starting gate at a speed of 250 km (150 miles) per hour, finding himself a second later 70 meters (76 yards) farther on.

Dr. Sound, true to his Gallic temperament, made a great deal of commotion, but penetrated a mere 340 meters (1100 feet) into the Strand, after breaking his own sound barrier.

The following participant, Mr. Brown, mimicking his own molecular motions, contrived within his allotted second to pass the gates of the Tower, fully 2 km (1.3 miles) from the start.

Next, the lady representative of the venerable firm of *Terra* (long-standing consultants in daily and yearly orbiting) reeled into the Square. Developing the speed of the daily rotation about her axis, she executed a prodigious bound which doubled the distance recorded by her forerunner.

Miss Ile summoned her accumulated experience of noisy escapes and managed a splendid burst to Croydon, 12 km (7.5 miles) away.

It was left to the diminutive *Alec Tron*, however, to dispose of Miss Ile's challenge. Whirling himself about an obliging atomic nucleus he flew off at a rare pace, arriving over St. Peter's in Rome just as "time" was called . . . a mere 2000 km (1200 miles), if you please!

Fully two thousand times as fast as the front-running Alec Tron, "Quick" Neutron was naturally regarded as a comic turn to round off the proceedings until . . . he vanished into thin air! The blank astonishment of all in the Square can be imagined when, shortly after his departure, a call from San Francisco announced that "Quick" had arrived over the Golden Gate Bridge precisely one-half second after his disappearance. Incredulous timekeepers calculated his trajectory: a fantastic 20,000 km (12,000 miles), which meant that "Quick" Neutron had outpaced the flamboyant Mr. Car by more than 140,000 times!

With an evil glint in his Grecian eye, Dr. Gamma sped quick as a flash from the starting gate, and was lost to view. He reappeared a little less than a second later at his point of departure, claiming that he had in the meantime finished seven round-the-world trips, a statement which turned out to be right.

2.15 Falling Light

Because of the basic interchangeability of energy and mass every photon carries with it mass also. Consequently photons, subjected to the Earth's gravitational attraction, will themselves fall toward the Earth. Experiments have confirmed that light is indeed bent out of its original course as it passes near planets of the solar system.

Light originating on the surface of the Earth is, of course, affected in the same way. For a large-scale demonstration, let us imagine a very powerful beacon projecting a beam of light "horizontally," that is, tangentially to the Earth's surface at the point of projection.

Disregarding purely optical effects such as refraction in the Earth's atmosphere, where is the light after traveling for 1/1000 second from the source?

Because of the Earth's curvature, it would be at a height of about 7 km (23,000 feet), 300 km from the beacon. But, in the time elapsed and the distance traveled, gravitational attraction would have deviated the beam 5/1000 of a millimeter toward the Earth's surface.

On this small time-scale the amount of the "gravitational" deviation looks very small, but if we could extend the experiment over a period of 8.5 minutes—the time needed to reach the sun—the light would fall 2.5 meters (8 feet).

And, Light from a Star . . .

When the light from a star reaches the Earth after passing through the immediate atmosphere of the sun, the angle of deviation induced in the path of the rays by the sun's gravitational attraction approximates to the angle subtended by a 2-meter ($6\frac{1}{2}$-foot) pole located 200 km (125 miles) from the point of reference.

2.16 Light-Pressure and Sound-Pressure [9]

Light from the noonday sun in the tropics exerts a pressure on the Earth's (or any flat) surface at that point of 2.5×10^{-4} dynes per square centimeter; this pressure, exerted over the whole earth at the same time would amount to a force of about 1,300,000 tons.

Taking an area equal to that of the United States, 9.4 million km^2 (3.6 million square miles), situated on the Equator, the force exerted by light-pressure, averaged over day and night, would be about 7500 tons, just more than the weight of the Eiffel Tower.

The force exerted by light-pressure would be like the weight of the Eiffel Tower . . .

Sound waves come to our ears as regular and rapid variations in air pressure which cause our eardrums to vibrate; this motion is communicated through a delicate mechanism of bones to our actual "hearing" organ where it is converted into a sensation of "sound."

For people with good hearing, the lowest limit of audible sound is produced by a pressure of 2×10^{-4} dynes per square centimeter, practically equal to the light-pressure exerted by the sun.

So, if we were to arrange apparatus to interrupt the sun's radiation intermittently, at a frequency in the audible range (say by a rotating wheel with suitable holes in it) we should be able to "hear" the sun's (light) pressure, provided that we could guard our ears against the scorching that might result from the exposure to the sun's rays for this purpose.

2.17 Seeing Red as Green

Professor Doppler found himself in court on the charge of having driven through a crossing against the red light, and attempted a defense on the lines of the well-known effect named after himself.

However, while the Bench were impressed with the erudition of the defendant, they thought it well to call in expert opinion. Dr. Speed the well known physicist, estimated that Professor Doppler must have approached the traffic light at about half the velocity of light to produce a shift in the apparent color of light from red to green. On hearing this the Bench pronounced a heavy prison sentence on the defendant on the ground of dangerous—and far too fast—driving.

2.18 β-Particles in Our Bodies

Every second about 1000 β-particles are released within the human body, the result of radioactive disintegration of the minute trace of the rare isotope of carbon with atomic weight 14, rather than the usual 12, which we all carry around with us.

2.19 A Win, Thanks to a Minute Superiority

Material containing hydrogen nuclei (spin 1/2), subjected to a strong magnetic field, 14 kilogauss exhibits resonance with a radiation of frequency 60 megahertz. This means that energy of the radiation is absorbed—and reradiated—when "spins," initially parallel with the applied magnetic field, snap over to the "antiparallel" state. But the remarkable feature is that in accordance with the Boltzmann distribution law the excess of parallel spins over the "antiparallels" is no more than 10 in a million. Even this minute excess, 1 in 100,000, suffices to give rise to the phenomenon of resonance.

2.20 Precision in Measuring

Distance

Experts in optical physics[10] can now measure the wavelength of light with a precision of 10^{-11} Angstrom units (Å) in 1 Å. If we could measure the distance through the earth to our antipodes with the same precision, the error would not exceed 0.1 mm.

Experts, by nature modest, do not always mention that this distance to our antipodes, would, if we had a land route throughout, require 5 days and nights of continuous driving at 100 km (60 miles) per hour (diameter of the earth: 12,740 km = 7912 miles).

That this degree of precision is not unique in the field of physical experiments is shown[65] by work done at Brookhaven, Long Island, where a synchrotron was used to verify Einstein's theory that signals of any kind cannot travel faster than the speed of light. By observing the effects of high energy pion-proton collisions, this theory was confirmed, even for distances less than 10^{-15} cm.

Traces

The presence or absence of some 10 or 15 elements in certain compounds in minute quantities can be a matter of life or death to human and indeed other life; typical of these are the iron in hemaglobin (see Section 8.22), and the iodine in thyroid gland hormone.

These minute quantities can, however, be detected and determined [66] by spectrographic analysis, among other newly developed techniques.

The electric currents which are to be measured in such analyses are so small that if we imagine the current needed to light an ordinary bulb as a tube large enough for the Earth to roll through, the current measured in spectrographic analysis would correspond to the thickness of a human hair, a ratio of 10^{11} to 1.

As another illustration, if all the atoms in 1 gram of lead could be imagined as pearls on a string, the resulting "necklace" would go 25,000 times round the Earth, a distance in all of 10^{9} km. But, using modern techniques for trace analysis, we can now detect and determine a quantity of lead corresponding to only 1/10 mm of this strange "necklace," that is a ratio of 10^{16} to 1.

A Comparison

In order to compare this result with that of the Brookhaven people, the scale of distances in the synchrotron experiment is imagined to be enlarged 10^{14} times, so that 1 cm can now be wound 25,000 times round the earth, like the string of (lead) pearls. On this scale the "synchrotron" measurements relate to a distance of 0.1 cm, and so may be said to be only one-tenth as accurate as the "lead" analysis, which related to 0.1 mm.

2.21 What Magnifying Power Means [11]

It was first suspected that there was a limit to the magnifying power of the optical microscope by Ernst Abbe, of Jena, in about 1900.

The best resolving power, or "resolution," attainable at that time was 600 Å, and the greatest magnification, 1000×. At this magnification, a man lying prone would appear to be 1.8 km (just over 1 mile) long, a distance equaled in New York by the trip from City Hall to Washington Square (or in London, from Buckingham Palace to Charing Cross).

In 1933, the Berlin scientist Ruska constructed a two-stage electron microscope with a resolution of 500 Å, and magnifying power of 10,000. This magnification would have made a man lying down seem as long as the distance in New York City from City Hall to the center of The Bronx (or from Buckingham Palace to Hampton Court), namely 18 km.

Today, electron microscopes are manufactured giving a magnification of 120,000, which is sufficient to show up an actual distance of only 10 Å as a visible mark on a photographic plate. By means of a binocular viewer the final magnification for the operator can be raised to 1.2 million, and on this scale our prone man would seem to stretch the whole distance from New York to Kansas City in the Middle West (or from London to Athens)—2100 km.

The very latest development is an ultrahigh resolution electron microscope, with magnification 2.5 million, enabling the operator to distinguish points separated, in reality, by only 2.3 Å. Applying the same magnification to a man, he would reach from New York to San Francisco (or from London to Teheran)—4500 km.

2.22 Weighing Letters

One hundred copies of this book contain a quantity of printers' ink which weighs about as much as the weight of one match.

This interesting equivalence was established with the help of an ultrasensitive balance, with which the gain in weight consequent on printing just 2 more letters in only one of the 100 copies could be measured and was found to be 5×10^{-9} gram.

2.23 Water—Boiling and Freezing [12]

The mathematical laws of probability tell us that there is a chance that in a tumbler of water—say 250 cm^3—all the fast-moving molecules will congregate in the upper part of the water, and all the slow-moving molecules in the lower part, with the result that warmer water will be at the top and cooler water at the bottom of the tumbler. However, this is an extremely rare happening; in figures, once in 10^{78} seconds.

In the limiting case, one might even have the extraordinary experience of seeing the water on top boiling while there was ice at the bottom of the tumbler.

However, as the Earth has existed, since solidification, for only 10^{17} seconds there is some time to wait before this separation into hot and cold can be expected.

Even if all the water on the Earth at the moment of its solidification had been put into tumblers—there would have been 10^{22} all told—probability laws foretell that the boiling-freezing phenomenon will occur only once in all this multitude of tumblers in a period which, by a remarkable chance, is the same number of years—10^{22}.

2.24 Weightlessness—an Everyday Phenomenon

Since every object which falls freely under the influence of gravity is in fact experiencing "weightlessness," this phenomenon is enjoyed by every youngster who—probably against parental orders—jumps off the pavement into the gutter.

On a larger scale, a high-diver, working from the 10-meter (33-foot) board, enjoys this remarkable condition for 1·4 seconds before striking the water.

2.25 Waves and Wrinkles

Whether wave motions are of a mechanical nature, and occurring in elastic media such as air or water, or of an electromagnetic nature (radiation), the wavelength, the frequency, and the speed of propagation of the wave (in the medium where applicable) are related by the formula:

$$V \text{ (propagation speed)} = n \text{ (frequency)} \times l \text{ (wavelength)}$$

This holds both for longitudinal and for transverse wave motion.

Although in practical situations, wavelength is the variable dependent

Table A
Wave motions characterized by wavelengths

Kind of wave motion	Wavelength		Spread in "octaves"
	Numerical units	Comparative basis	
The shortest wavelength	10^{-4}—10^{-1} Å	Electrons and atomic nuclei	10
X-rays	0.1–10 Å	Atoms	7
Ultraviolet light	10–4000 Å	Molecules	9
Visible light	4000–7000 Å	Smallpox virus	1
Infrared radiation	7000 Å–0.5 cm	Between *Escherichia coli* and a pea	14
Radar	0.5–50 cm	Between a pea and a cucumber	7
Sound	2 cm–20 meters	Between the breadth of a piano key and the orchestra's platform	10
Water waves	2 cm–200 meters	Between the thickness of a stout cane and the length of an ocean-going vessel	13
Radio waves Short (for communication with artificial satellite)	1 cm–50 cm	The satellite itself	12
Medium	50 meters–500 meters	Between the height of a radio-station aerial and the length of the street in which the station is located	3
Long	500 meters–1000 km	Between the length of the street and the distance from London to Nice, or from New York to Chicago	11
Electrically-induced heat	10 meters–25,000 km	Between the width of a house and about the distance from the North Pole to the South Pole, measured along the Earth's surface	21
Electrical waves in high-tension "power" lines	500 km–25,000 km	Between the distance from London to Cologne or from New York to Pittsburgh, Pennsylvania, and the distance from Pole to Pole, as above.	6
Total			75

Of these 75 octaves, men can only apprehend the 25 or so octaves from visible light, through heat, to sound and water waves directly through his natural "senses"; for the remainder he needs apparatus of one sort or another, sometimes of great complexity.

on the propagation speed in the medium, rather than frequency which usually comes from the means of generating the wave motion, we shall use wavelength as the basis of comparison of a number of wave motions; in this way the waves become, in a sense, visible.

Taking a hint from the study of sound, we will express the width of the wave band covered by any particular radiation or wave motion in terms of "octaves"; each octave means, in a given medium, the doubling of the frequency or the halving of the wavelength (Table A). Moreover, for sound waves we assume air as the medium, for "water waves" of course water, and for the rest, "vacuum."

We find that in the frequency range significant for man and his doings there is a range of wavelengths from 10^{-4} Å to 25,000 km (in frequency terms, 10^{22} to 30) which means 75 "octaves":

2.26 The Fly on the Liner

Whenever a fly alights on an ocean liner of about 35,000 tons, the ship tends to sink lower in the water by one-tenth the thickness of an atom (0·1 Å)—this can be measured at present by means of an electrostatic capacity meter.

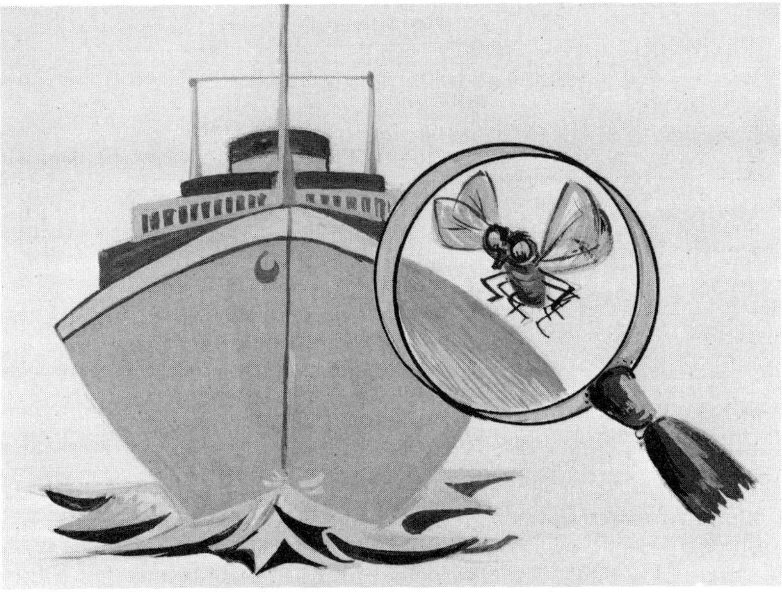

. . . the ship tends to sink by one-tenth the thickness of an atom.

If the fly lands on the handrail, say 15 meters (17 yards) from the center line of the ship, the resulting downward deflection of the ship on the same side will be about twenty times greater (unless the vessel is efficiently stabilized). In fact, it is not even necessary for the fly to touch the ship at all. If it merely hovers just above the deck, the vertical pressure of the airstream generated by its wings will have practically the same effect on the ship.

2.27 Grandfather's Clock Near to Perpetuum Mobile

A pendulum, in the course of a single swing, converts potential into kinetic energy and back again into potential energy.

This double conversion is accomplished in a good grandfather's clock with only tiny energy losses, resulting from airdrag and friction in the suspension.

The result is that a spring, weighing 50 grains (about 2 ounces) can swing a 100-gram (4 ounces) clock pendulum back and forth for a week. It would take a spring, weighing as much as a couple of men (together 140 kg = 300 lb), to do the same job, if the back and forth motion were entirely horizontal and wasted all its energy at the end of each stroke[67].

2.28 An Examination in Physics [51]

An unusually acute examination candidate, faced with the following problem: "Demonstrate a connection between:

(a) A large modern electricity generating station
(b) An electric shaver
(c) A regrettably sluggish official
(d) World War III,"

was able to devise this ingenious answer:

"(a) It is now possible to make X-ray photographs of very rapidly moving objects, but electric power is required at the rate of about 3000 amperes at 100 kilovolts, thus at 3×10^9 watts, the output of a fair-sized power station.

(b) On the other hand, exposure times are exceedingly brief in this technique, about 10^{-7} seconds; so the electricity used for one exposure is 100 watt-seconds, which would run an electric shaver for half-a-minute.

(c) In the laboratories where this sort of work is done, 10 exposures per day would be a normal performance, and this means actual use of the equipment for 10^{-6} seconds. Since there are approximately 10^5 seconds in a day, this means a time-efficiency of about 10^{-9} %, and even the slowest office type ever encountered was doing better than this.

(d) The cost of one exposure comes to about \$3.00, including a depreciation allowance based on the time of actual use; even so the rate of expenditure amounts to \$30 million per second, easily enough to cover the cost of present-day war."

Chapter 3

The Building Blocks: Atoms and Molecules

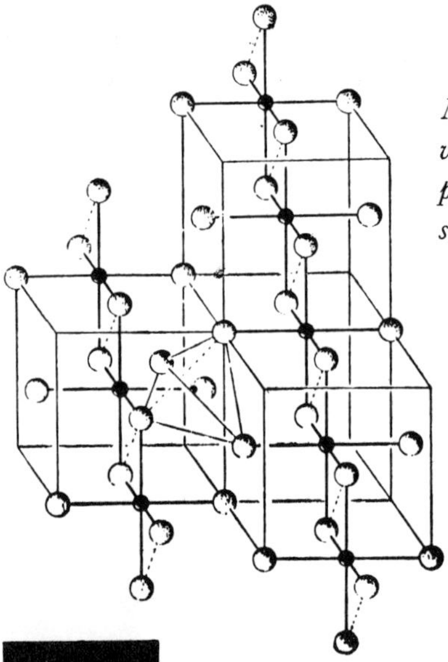

Might not things external to ourselves vibrate in union with our moods and passions, atom to atom in love of strange affinity?
(Oscar Wilde)

3.1 Molecules, Small as They Are

Most molecules are of such small dimensions* that even scientists experience considerable difficulty in visualizing their relationship to the everyday physical world.

● *A Teaspoonful of Molecules*

One teaspoonful of water 1 cm^3 contains at least as many molecules as the Atlantic Ocean contains teaspoonfuls of water.

● *Poseidon's Water Molecules*

If Poseidon, or any other person, living long enough ago, had emptied

* The number of molecules in 1 cm^3 of water is about 3.3×10^{22}.

a pint (about half a liter) of water in the sea—and assuming that this water were now evenly mixed through the seas of the world—then each pint of water taken at any source or tap, would now contain up to 5000 molecules of the original contents of Poseidon's pint.

• *A Monolayer of Molecules*
If every square meter of the Earth's dry surface were occupied by six men, their total would approximate the number of molecules in a monolayer of air 1 cm^2 (10^9 million).

• *A Handful of Snow*
If all the molecules in a handful of snow were magnified to the size of a pea, there would be snow enough to blanket the entire surface of the Earth to a thickness covering the Eiffel Tower or the Rockefeller Center in New York, i.e. 300 meters (330 yards).

3.2 Macromolecules

The 200,000 different kinds of plants and the million distinct sorts of animals which are so far known to have evolved, representing the flora and fauna around us, are for the most part composed of macromolecules.

If the nucleus of the carbon atom, the basic constituent of these macromolecules were to be represented by an apple pip which was placed in the center of a large stadium, the electrons on the outer shell would occupy orbits as far from the center as the spectators in the rearmost seats.

A stretched-out macromolecule of rubber incorporating carbon atoms on the same scale would cover the distance between London and Singapore (or New Orleans and Cape Horn), 10,000 km (6000 miles).

A molecule of DNA (see Section 8.5) on this scale would stretch from the North to the South Pole (20,000 km = 12,500 miles) and have a width of 2.5 km (1.6 miles).

A man, elongated in the same ratio, standing on the Earth in the plane of its orbit would be a giant so huge that the Sun would be only halfway up his shins.

Diameter stadium = 150 meters.

. . . halfway up his shins.

3.3 Compressing the Empty Atoms

If one could enlarge the nucleus of a carbon atom to the size of one of the airports of New York or of Paris, the electrons in the outer shell of the atom could be thought of as small sports planes circling around over Gander and the Bermuda Isles or Oslo and Algiers.

The weight of 1 cm^3 of such closely packed nuclear material would equal the freight carried by 1000 trains, each of 150 20-ton wagons; in short, 3 million tons.

Another way of expressing the density of nuclear material would be to point out that the remarkable cubic centimeter mentioned, would exert such a gravitational force on the hands of an experimentalist trying to work with it that in falling through the floor—which it would of course do even if the floor were made of the best armor plate possible—it would tear the hands off the experimentalist (see also Section 2.12).

3.4 Expectation of a Change in the Structure of a Molecule [13]

One way of expressing Boltzmann's "distribution law" is:

$$t = \tau^{W/kt}$$

In this formula t expresses the time interval between successive occurrences of a structural change when bonds of energy content W are disrupted; kT expresses the Boltzmann gas constant, i.e., the thermal energy of the molecules; τ is a constant.

It is amazing how relatively minor changes in the ratio of W to kT result in enormous differences in t, the time of expectation of the structural change. With $W/kT = 30$, t will be only about 0·1 second; but with $W/kT = 50$, t becomes 16 months, and with $W/kT = 60$, $t = 30{,}000$ years!

3.5 Concerning Electrons

● *Small as They Are*

Inconceivably small though individual molecules may appear to us (see Section 3.1), they are still of vast proportions when compared to electrons.

If, after the Flood,* Noah had set himself to string electrons on a thread at the rate of one a second for 8 hours a day, the chain so formed would today still be only two-tenths of a millimeter long! Composed of hydrogen atoms, the chain would be something like 25 meters (80 feet) by now.

● *Moving Fast or Slowly*

In hydrogen the electrons travel along their orbital paths with a velocity of 2000 km per second, i.e., at a speed about one hundred times

* Supposed to have taken place around 3000 B.C.

as great as that attained by any rocket or missile yet launched by man, but light travels a good hundred times as fast as the electrons.

In a conductor, however, their progress may be very slow indeed. For example, when a current of 1 ampere flows through a copper wire with a cross-sectional area of 1 mm^2, the speed of the electrons is as if a walking man would cover 6 meters (6.6 yards) per 24 hours. Their forward motion is then like that of a very viscous fluid pressed through a narrow-gauge tube; it is only the tremendous number of electrons in transit which results in the transfer of an appreciable amount of energy.

Consider now a 100-watt lamp burning for 1 second at 100 volts. If the displaced electrons were to be represented by an army of soldiers marching shoulder to shoulder in ranks half a meter apart, each rank would contain the population of the Benelux countries (22 million) and the front and rear ranks of the column would be separated by the distance between the Earth and the sun (about 150 million km or 93 million miles)!

3.6 Hitting Nuclear Targets

● A neutron flux of 10^{11} neutrons per second per square centimeter signifies that the atoms of irradiated material would suffer collisions at the rate of 10^{11} per second per cubic centimeter.

Although every "hit" by a fast neutron results in a few hundred displacements of atoms, in a whole year only 1 atom in 300 of an irradiated solid would have undergone displacement.

● Or, expressed in another way:

In the process of uranium fission, the chance of a single quick neutron hitting and splitting a uranium nucleus can be expressed in the following way.

If we represent the diameter of the uranium atom by the ground-level cross section of a giant California redwood tree (10 meters), the uranium nucleus will resemble a pinhead (1 mm across) in the heart of this tree: the quick neutrons will be 7/100 of the size of this nucleus.

Consequently, most of these neutrons pass through a vast forest of redwoods (uranium atoms) before striking a pinhead (uranium nucleus). In fact, on our chosen scale, the mighty forest of (closely packed) redwoods would cover twenty times the distance between the Earth and the moon (i.e., $20 \times 350{,}000$ km).

● And for neutrinos:

Small as the probability is that neutrons will hit a target, it is enormously greater than the chance of a neutrino striking an atomic nucleus. A neutrino with energy level a few million electron volts could penetrate more than 100 light-years of liquid hydrogen before interacting with a hydrogen nucleus.

3.7 The Old-Fashioned Hairsplitters

Recently an ultrafine cutter has been devised of which the blade is only some 10 atoms thick (2×10^{-7} cm), producing sections of the same thickness. These are 10 times as thin as the wall of a soap bubble and 100,000 times as thin as the average human hair.

Those who split atoms are therefore still 10 times as clever as the microtome cutters and at least some million times as clever as those who split hairs.

3.8 The Classic Pith-Ball Experiment [4]

The rather feeble electrostatic forces shown by the elementary classic experiment with pith balls leaves one unprepared to expect that the strength and rigidity of steel should be due to identical forces.

The charge on an electron is only $1{\cdot}6\times10^{-19}$ coulombs, but a single electron may be within 10^{-8} cm of some nuclei in a solid, all pulling at it. The intense square law of force makes these pulls of impressive size for the tiny particles involved.

The weak forces acting in the pith-ball experiment arise from the fact that the excess or deficit of electrons on each ball is exceedingly slight. It would probably be less than 1 electron per 100,000 million atoms.

If a pair of pith balls had one deficit and the other an excess of as much as 1 electron per 10,000 atoms, they would attract each other with a force of a thousand tons from a meter apart and, if released, would strike each other with an energy equivalent to the explosion of about 10 tons of TNT.

3.9 Tap, Tap, Tap . . . [63]

In a gas under "ordinary" conditions, as in the air around us, any one molecule will collide, on the average, with 10^{12} other molecules in each second or even more. A bond between two water molecules breaks and rebuilds itself at about the same rate.

Now think of a man tapping the table with his finger. If he is very dexterous he may manage 10 taps per second, and, assuming that he could keep this up continuously, he would have to have begun his tapping at the time of David and Goliath—about 1000 years B.C.—in order to equal the number of collisions experienced by a molecule of the air or the number of breaks of a bond in the last single second of this period.

3.10 The "Traffic Jam" in Water

Since in water the molecules are rather tightly packed together, the mean free path of a molecule between successive collisions is only 2.5 Å, that is a distance of the same order as that of the molecular diameter.

In spite of this close packing, the molecules of water move, at room temperature, at about 10 times the speed of a modern jet aircraft, thus at 7000 km (4400 miles) per hour, from which it is clear that the number of collisions per second runs into millions and that the molecule which we are watching will change its direction as many times.

Hampered by this traffic jam, the effective speed of our molecule "over the ground"—which means, in fact, the diffusion velocity—is only about the speed attained by a frog, aiming to cover 100 meters (330 feet) in an hour. And, we may guess, alternately jumping and walking to achieve this end.

3.11 The Elusive H^+ Ion

Although we often speak and write of "the hydrogen ion," in its usual surroundings, a solution in water, it actually exists almost entirely in the hydrated form, associated with a molecule of water, H_2O, thus in fact "H_3O^+." Indeed, the equilibrium lies so far in favor of H_3O^+ as against H^+, that only one true H^+ ion would be found in a volume of water equivalent to 1,000,000 times that of the presently known universe.

3.12 A Material Decomposes [14]

Ethylbenzene is a widely used reagent or component of which the routes and velocities of decomposition are fairly well known. Let's consider a bottle of it, stored in initially pure condition on some odd planet 200 times as long ago as the Earth's beginnings, that is 10^{12} years ago, but kept till now at today's "room" temperature.

By now, for every 3×10^9 molecules of ethylbenzene originally present, there would be just one molecule of the most abundant decomposition product. But to find it would be as difficult as picking out one particular shrimp on a needle from the whole area of the Earth's seas and oceans.

3.13 Saturating the Oceans [14]

You don't need to be a large-scale financier, or even unusually generous, to saturate the Earth's oceans with (a) salt: 12 grams of lead sulfide—price less than a shilling, or a few cents—is all that the whole of the sea water on the Earth can dissolve. Having made your gift, in each liter of water there will be only 60 molecules of it.

Even less soluble in water is the red sulfide of mercury, also called "cinnabar": to make sure that you have captured one molecule of this unusual material you should take 150 liters of a saturated solution, which is easily at hand, in the oceans.

3.14 Where are the Largest Numbers, in Astronomy or in Chemistry?

The immense distances between the planets, stars, and other heavenly bodies, and the vast number of stars, make it natural for us to call any unusually large number "astronomical" in common speech, even when the number has really nothing to do with the science of astronomy.

What are we actually implying, in using this word?

Accepting that the diameter of the perceptible universe is 10,000 million light-years (see Section 4.4) it is clear that the description "astronomically great" must apply to very great distances indeed, the more so, seeing that a light-year amounts to 10^{13} km. The significance of this "light-year" is made clear in Section 1.1.

However, astronomers do not hold a monopoly in very large numbers, as appears from the comparison of the astronomers' number of heavenly bodies with the chemists' numbers of molecules.

Accepting the further estimate that there are in the perceptible universe 100,000 million galaxies each comprising about 100,000 million heavenly bodies, we find that the total number of heavenly bodies is about 10^{22}, a number which strongly supports the use of the word "astronomical" to denote "a very great many."

But we must realize that even this number is minuscule compared with some numbers with which we are confronted every day. The housewife washing the breakfast crockery throws away a good 10^{22} molecules in every thimbleful of water that goes down the drain. Thus in 1 cm^3 of water there are more units than in the astronomers' known universe, not to mention the numbers in that measure more familiar to the housewife, the "jugful."

3.15 Disappointments in Mammoth Research [5]

The search for active chemicals with some specific property, for example for antimalarial drugs, often involves the syntheses of whole series of organic compounds, differing only slightly from each other, which are then tried out—"screened," in modern terms—on small animals such as mice.

The antimalarial drugs are indeed probably the best-known example of this sort of technique. Before Plasmoquine was discovered in 1924, about 10,000 possible compounds had been synthesized and tested, and for Atabrine the number was even more than 12,000. Even these compounds left so much to be desired that during World War II further research was undertaken in the United States in which about 50 research institutes and laboratories cooperated.

The outcome, after examination of about 14,000 "possibles," was Chloroquine, which, however, turned out to be identical with a compound already synthesized in Germany in 1934, but at that time not correctly appraised as to its therapeutic value.—Disappointment No. 1

But even such mammoth investigations are only minuscule compared with what might be undertaken in the field of organic chemistry.

A chemical manufacturer had a plan to investigate the influence and effect of eight different substituents,

fluorine	hydroxyl(OH)
chlorine	hydrosulfide(SH)
bromine	methyl group(CH_3)
iodine	phenyl group(C_6H_5)

when introduced into a complex polycyclic molecule with this basic structure:

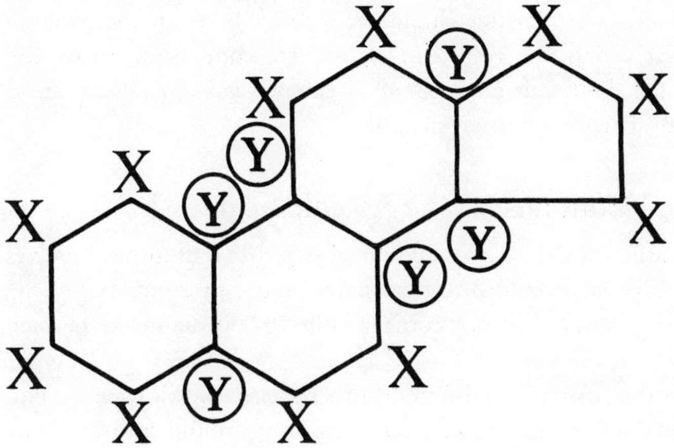

Polycyclic Molecule.

In this there are 11 positions (X) where two substituents, and 6 positions (Y) where one substituent can be introduced.

Before the work was really under way, a statistician discovered that to get only 1-gram samples of every possible compound in this interesting series, 10^{18} tons of material would be needed.

Supposing that the specific gravity of the samples was around 1, a stack of drums about 100 meters (about 330 feet) would be needed, based on an area 50 times that of the Earth's total land surface!

In view of limited storage capacity, the project had to be given up.—Disappointment No. 2

3.16 Inside an Internal Combustion Engine

Internal combustion engineers face the difficulty [15] that in the almost instantaneous explosion, lasting say 1/100 second, there is not time for the reaction partners, gasoline (petrol) and oxygen, to come fully into contact with each other to ensure optimum performance. Indeed, for stoichiometric proportions 1 molecule of petrol (gasoline) must be surrounded by 50 molecules of air, of which only 10 are oxygen. For this reason combustion can be speeded up by supplying an excess of gasoline which gives a better chance of contact between the reaction partners, but which also results in incomplete combustion to carbon monoxide, the poisonous component of exhaust gases, rather than to the relatively harmless carbon dioxide.

3.17 Social Contacts on the Atomic Scale [4]

Interpenetration of the electron clouds of two different atoms involves strong repulsive forces and in the ordinary course of events the nuclei of neighboring atoms will not come within 10,000 diameters of each other.

Although the reasons are different, one can see that, if each human being maintained a number of stones whirling around him at a fair fraction of the speed of light in unpredictable orbits on mile-long strings, social contacts would be rare.

3.18 Odd Bits

- The usual range of energies of nuclear interactions approaches a million times the usual range of energies involved in chemical reactions.
- The tiny proportion of tritium ($^{3}_{1}H$) formed in rainwater by the action of cosmic radiation has a half-life period of 12 years; in 1 liter (or about 1 quart) of rainwater, the average number of atoms of tritium which break down in each minute is 7.
- The number of free electrons in 1 cm^3 of a conducting metal, which constitute the "electronic cloud" whereby conduction of electricity can occur, can be of the same order as the number of molecules in a cubic centimeter of liquid, that is about 10^{22}.
- In a crystal of ice just below the melting point, every molecule normally rotates at the rate of 100,000 times per second. That is more than 100 times faster than the rotor of an ultramodern centrifuge (see Section 8.8).

Chapter 4
Space

Matter is ripples in a non-existing ether.
(M. Vertregt)

4.1 The Universe [72]

● *The Constituents of the Universe*

Present-day cosmology suggests [17] that the most abundant material in the universe, organized in the form of atoms and molecules, is hydrogen which seems to comprise, in interstellar space, about 10,000 atoms per cubic centimeter, and to exist even in intergalactic space in a proportion of 1 atom per 10 cm^3; on the Earth 10 cm^3 contain some 10^{20} atoms.

By far the greater part of all matter in the universe consists of free particles: protons, neutrons, electrons, γ-rays. The main portion is still in the fourth state of aggregation as plasma in the interior of the stars.

Even including visible objects, it seems that hydrogen, albeit the most tenuous material, constitutes 93% of all atoms in our Galaxy with helium accounting for nearly all of the remaining 7%, so that in spite of its lightness, hydrogen accounts for about 76% of the total volume of our Galaxy.

All the other elements together account for but a small fraction of 1% of all atoms, so that iron, oxygen, silicon, and so on, familiar to us on the Earth, are really rare elements in the universe, and those which we ordinarily term "rare" elements, such as gold or platinum, are extremely rare indeed in a cosmic sense.

● *Some Extremes of Density*

Intergalactic space: 1 atom in 10 cm^3
Interstellar space: 100,000 atoms in 10 cm^3
The Earth's atmosphere: 10^{20} atoms in 10 cm^3
(near the surface)

Now in grams:

The Earth's solid surface: 55 grams (10^{24} atoms) in 10 cm^3
In a "dwarf" (condensed) star: 10^{15} grams (the weight carried in 100 million railway wagons) in 10 cm^3

● *The Origin of the Universe*

It is generally agreed that the red shift (Doppler effect, applied to light) in the spectra of stars and nebulae means that all these bodies are receding from us (and from each other) at speeds which are found to be greater the more distant the object and which are, for the most distant objects, not far removed from the speed of light itself.

Two distinct theories [16] have been developed to fit this observation into some kind of physical and temporal continuum. The so-called "big bang" hypothesis suggests that at some specific time the explosion of one primeval "super-atom" occurred, the fragments from which now constitute the known, and expanding, universe.

On a quite different basis, the "steady state" hypothesis predicates the continuous, though very slow and rare (by terrestrial standards) appearance of hydrogen in intergalactic space. We may term this "creation" but we may also relate the phenomenon to the transformation of energy into matter (see Section 4.2). To introduce some numbers, the rate of hydrogen formation is estimated to be 1 atom per year in the volume occupied by the Empire State building, or, in Europe, Saint Paul's Cathedral in London; but even this "tiny" event represents the "creation" of no less than 10^{32} tons per second throughout the entire universe.

The gradual agglomeration of these stray hydrogen atoms into matter is assumed to result in a pressure which is relieved by the observed outward movement of all heavenly bodies and the end result is a universe uniform in both space and time.

We have also that modification of the "big bang" idea which includes the suggestion that after a primary explosion a fairly uniform distribution of hydrogen gas became established, but that an influence termed "cosmical repulsion" then came into play, causing agglomeration of material into galaxies and stars, as well as the general expansion. After about 40 thousand million years it seems that gravitational forces may overcome the cosmic repulsion and contraction begin, to be succeeded by expansion again in a never-ending cycle of period about 80 thousand million years. But the ordinary man can not feel himself much affected even by this theory, for the present expansion is believed to have begun "only" about 10 thousand million years ago, so the next reversal is not due for another 30 thousand million years—

4.2 Conversion of Radiation into Matter [18]

It is estimated that in the perceptible universe mass is distributed, overall, at a density whereby each gram has 30 times the volume of the Earth at its disposal.

Einstein's formula, $E = mc^2$, enables us to express the radiation energy as an "equivalent mass." This leads to the conclusion that this "equivalent mass" in the universe is so much less dense than the "mass" itself that for each gram of the equivalent mass a volume 3 million times that of the Earth is available.

Thus, so calculated, energy present as radiation is a very small proportion. Expressed in terms of power present in a given volume it is something like the power of a small flash lamp compared with that of an automobile engine: 1 to 100,000.

It would fit in with the steady state theory of the universe that we are now seeing net transfer from mass to radiation, which at some far future time will be reversed to the direction which it had also in the past.

It is estimated that the time cycle of this self-regulating system is 1.3×10^{15} years, in which all mass will have been transmuted to radiation and back again, and therefore renewed. In other words, when the Earth has reached 300,000 times its present age, its whole mass will have been renewed, with radiation as an intermediate state.

The reader is advised to think over the following definition by Fritz Kahn: "Material" is that part of the "mass-energy continuum" which reveals itself as "mass." "Space" is that part of the continuum which is not revealed as "mass" but where it is only demonstrable as "energy".

Radius of the perceptible universe...	10^{10} light-years (10^{23} km)
Density of mass in the universe ...	3×10^{-29} grams per cubic centimeter
Equivalent density of radiation in the universe...	3×10^{-34} as grams per cubic centimeter
Age of the Earth	$3–5 \times 10^9$ years

4.3 Our Solar System on an Earthly Scale

No schoolroom map is large enough to show on the one hand the distances from the sun to the planets and at the same time the sizes of the planets themselves. The latter would just be tiny fractions of a mark to be seen with a microscope.

To get a clearer picture, let us use the Earth's whole surface as our astronomical chart, with a scale in which 1 centimeter (cm) = 1 kilometer (km), or 1 inch = about $1\frac{1}{2}$ miles, or $1{:}10^5$:

We need an airplane to travel about on this map; suppose that we set out from the southernmost point of Manhattan Island, New York, observed by the Statue of Liberty.

We set out from the southernmost part of Manhattan.

While still above New York City, in fact over Washington Square, we see the *moon* depicted at about the size of our airplane (35 meters, about 100 feet across) on the "Earth"-map; we have traveled only 3.8 km (2.4 miles) but this represents 380,000 km, the real distance from the Earth to the moon.

Now we must fly to a point just beyond Chicago, a total of 1500 km (930 miles) from our starting point, before we "pass" the *sun*, a gigantic globe with its diameter equal to the length—14 km (9 miles)—of the District of Columbia, the "Federal" area around the city of Washington.

We will assume that on our Earth-map all the planets—including our Earth—are shown along a straight line, and not in real positions in their orbits; and so en route to the sun we shall have passed *Venus*, 400 km (250 miles) from the start, and about 123 meters across (400 feet) so about "fitting" the Hudson River at New York, and also *Mercury*, 900 km (570 miles) from the start, 50 meters (160 feet) across, so like a large aircraft.

Now continuing from Chicago, we fly almost to San Francisco before finding *Mars* on our map, 3700 km (2300 miles) from the starting point which will be, on our chosen scale, 67 meters (220 feet) in diameter, so corresponding in size with a large supersonic "line" aircraft.

But after this we have a really long "hop"—for we cannot concern ourselves just now with the asteroids, the swarm of tiny planets between Mars and Jupiter—to a point well past the Hawaiian Islands, 9000 km (5600 miles) from our start at New York. Here we pass the giant planet *Jupiter*, with its diameter of 1400 meters (not far short of 1 mile).

Still another long flight is before us, for our next planet, *Saturn*, will be "found" well into Australia, 16,000 km (10,000 miles) from the start. This is also a giant—even disregarding the rings, its diameter is the equivalent of 1150 meters on our Earth-map (about $\frac{3}{4}$ of a mile).

Continuing our westbound flight we have still to see two more planets, about the same size, 500 meters (1650 feet) across, or the size of the lawns around the Pentagon Building at Washington, D.C.

The first of these, *Uranus*, will be found at the Congo, in Africa, 30,000 km (18,600 miles) from New York; but we shall have to continue around the Earth and pass New York and San Francisco for a second time before we find *Neptune* in the neighborhood of the Hawaiian Islands, distant 47,000 km (29,000 miles) from the starting point.

Finally, after passing Australia for a second time, we find the last-discovered planet *Pluto*. Fortunately it is shown at its mean distance from the sun, otherwise its very elliptical orbit might require us to travel much further; but our meeting has taken place 60,000 km from the start, that is, 37,000 miles, after we have circled the Earth $1\frac{1}{2}$ times.

4.4 The Universe on an Earthly Scale

We have just seen (in Section 4.3) how our solar system would appear if charted on the surface of the Earth itself; let us now consider how the universe would appear, charted in the same way.

For this we choose the scale 1 cm = 1 light-year (10^{13} km, or 0.6×10^{13} miles); this is equivalent to $1{:}10^{18}$ (instead of $1{:}10^{5}$ on the previous scale).

Once again we set off by airplane from the southernmost point of Manhattan Island, in New York, and soon we are clear of our *Milky Way Galaxy*, which appears as a cloud of stars about 1 km (0.6 mile) in diameter, centered over the Battery Park and New York harbor, some 200 meters (660 feet) thick at the center tapering to about 25 meters (75 feet) near the edge.

Looking back with a powerful telescope we might be able to discern our *solar system* as a tiny object 1/100 the size of a pinhead, about 200 meters (220 yards) from the outer edge of the cloud; and within this our Earth is a microscopic speck.

This sort of comparison helps us to realize clearly how small our solar system is compared with the dimensions of the known universe; in numbers, the mean distance from the sun to the outermost planet, *Pluto*, is about 5 light-hours (not light-years!) thus a very small fraction of our unit of scale, the light-year, which is represented by 1 cm.

Looking around again from our aircraft, we would see at about 240th Street, 20 km from our starting point, the *Andromeda Nebula*, which is a galaxy of about the same size as our "Milky Way," which we have just left.

Continuing our flight westward, we see galaxies, some larger, some smaller, every 10 km or so; but over the Pacific Ocean, beyond San Francisco, we reach a point that a few years ago was the furthest range of the most powerful telescope at the Mt. Palomar observatory. This point actually lies 2000 km (1250 miles) beyond the Hawaiian Islands, so 10,000 km (6250 miles) from our starting point at New York, representing thousand million (10^9) light-years. Present equipment has "enlarged" this observable universe to 10^{10} light-years: on our map 2.5 times around the Earth.

Compare this enormous distance with the size of the Earth, hidden as a minute part of an object itself 1/100 the size of a pinhead situated at the starting point of our journey.

4.5 Journey into Space with Doppler

A spaceship sets off, so powered that its speed continuously increased at 3 meters per second^2 ($=0.3\ g$), an acceleration which was still within the endurance of the specially trained crew.

The physicist on board, Professor Doppler, told them that the increase in speed would have some remarkable results concerning the colors of the stars which they would see. As a matter of fact the following was observed:

● The crew members looking forward saw in the course of time red stars (wavelength 7000 Å) shift to yellow, yellow ones to green, green stars to blue, and blue ones to violet (4000 Å). After 8 months—the vehicle was then moving at 20% of the speed of light—the blue stars, seen as violet, disappeared from sight. The originally red ones disappeared after 17 months, when the speed of the vehicle was 45% of light velocity.

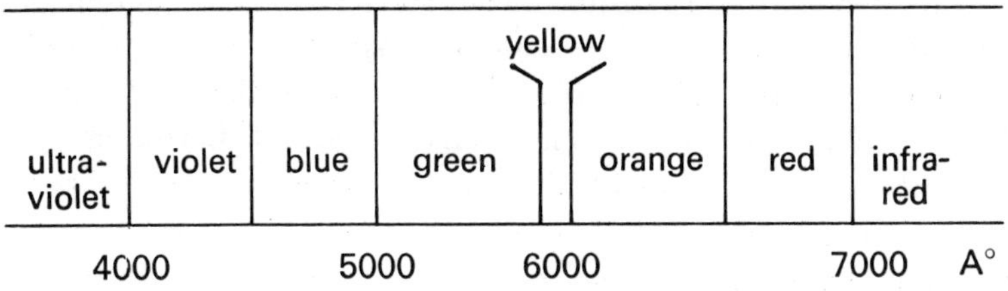

The observed spectra.

It was fortunate that Professor Doppler had remembered to include a few insects in the flight baggage, since he planned to make use of the ability of some species to make observations in wavelength ranges, inaccessible to man's ordinary senses. Indeed, it appeared that a honeybee still picked up radiations of the "lost" stars in the ultraviolet at 3000 Å and a water flea did so even at wavelengths of 2200 Å.

● The crew also observed that an initially invisible star detected during the flight due to its heat radiation (infrared, 8000 Å), appeared during the trip in the sky as red. Then it followed the same course as already mentioned for red stars: a generally accepted feature at that stage of the journey.

The men then began to observe straight ahead a black spot, increasing day after day, until no more stars were visible in the direction of the bow.

● Looking backwards the crew observed changes in exactly the opposite sequence.

First the water flea made some motions, indicating that something was happening in the 2200-Å range; then the honeybee became unquiet due to its observations in the 3000-Å area. Some time later the crew noticed a blue star which gradually shifted via green and yellow to red and then disappeared into the infrared region, about 2 years after the departure of the vehicle. Then a black spot was observed in the backward direction, which also increased day by day until nothing remained visible to those looking from the stern.

● Professor Doppler's assistant, not so well trained in physical matters, forecast that when the speed of the ship would approach light velocity, the crew would need a leaden shield against the radiation from certain stars which would then have the character of X-rays with the effective wavelength only 10 Å. Even this shield would have to be continually thickened so as to keep out the "gamma" rays of 10^{-2} Å wavelength, the form in which the light would soon be received at the spaceship.

Taking at that moment a nostalgic glance toward the Earth, the crew would only keep in touch with their mother-planet by use of a radio receiver, working far in the infrared.

Professor Doppler pointed out, however, that at such speeds perhaps new effects, connected with time dilatation may appear. But it seemed to him that apart from what we see, the actual frequency of all observed radiations has in all cases remained the same and so also the emitted photoenergy. Consequently he did not believe that any shielding against X-rays or any rays would be necessary even at these high speeds.

4.6 New Windows on Space

Some 40 years ago, the Milky Way, the vast system of stars of which our sun is one quite ordinary member, could be investigated only by optical means. Because of absorption of light in the thin dust clouds which are found among the stars which constitute the Milky Way, only a small part (as is now known) was perceptible to astronomers; in fact, the region out to about 5000 light-years.

Radio waves, however, are not absorbed by these dust clouds, and so reach the Earth from all parts of the Milky Way. Because of this, the new technique of "radio astronomy" has made investigation of hitherto quite inaccessible regions of space possible; in fact, we have been given a new "window" out into space.

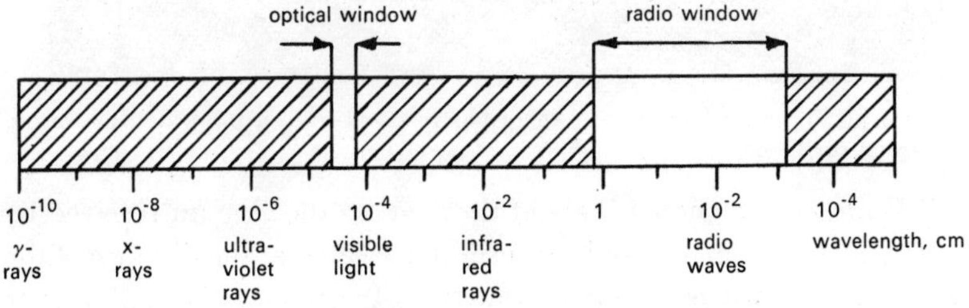

. . . the two present windows out into space.

In 1944 it was suggested [20] that the Milky Way should be emitting photons of about 21 cm wavelength, consequent on the transition of some of the electrons in hydrogen atoms to a lower-energy state. In 1950 this was confirmed by observation; but statistically it was estimated that of 100 hydrogen atoms for which the transition was possible, 50 would remain unchanged after 11 million years.

Notwithstanding the minute probability of the change, and notwithstanding the tiny amount of energy of the photon, 10^{-17} erg, the absolute number of hydrogen atoms involved in the transformation is so large that on the average the energy reaching the Earth from this source is sufficient to operate a small light bulb—3 watts.

Of this 3 watts, a radio telescope with an area of even 250 square meters will only pick up 10^{-17} watts; however this is detectable with modern apparatus.

. . . each picking up an amount of energy as received from a 0·5-watt transmitter, located on the moon.

This amount of energy is about the same as the amount received by a television set from a 100-kW transmitter located at the distance of the sun, or a 0.5-watt transmitter located on the moon. This is however about 1 million times too small to produce any effect in an ordinary television receiver; to get a perceptible signal, we should need an aerial of 1 km^2, about the area of London's Kensington Gardens.

À Propos

The detection of this 21-cm wavelength radiation (now known as the "Van de Hulst noise") by the Harvard student H. Ewen, using a radio telescope of his own construction, must be one of the least costly discoveries on record that helped to unlock the secrets of the universe, for the cost of Ewen's instrument was less than $500.

4.7 Counting the Stars

The stars which we see as separate points, whether with the naked eye or with telescopes, are situated in "our" Galaxy which, it appears, has the shape of a disc with a diameter of 100,000 light-years and a thickness varying from 20,000 light-years near the center down to about 1000 light-years as the edge is approached.

The number of stars in this Galaxy is estimated at 10^{11}, a number which provides 30 stars for each inhabitant of the Earth. But we can see only a quite small proportion of these, mainly because visibility is restricted by the presence of great clouds of interstellar dust.

By way of illustration:

- We can see with the naked eye a number of stars about equal to the population of a small provincial town, of say 2000 persons; counting these stars at the rate of one per second would occupy us for half an hour.
- With a small telescope, we can see a number of stars about equal to the population of Brussels—1 million.
- With our most powerful telescopes, we can see a number of stars about equal to one-third of the Earth's population, thus about 1000 million. Counting this multitude at the rate of one per second would occupy our time every night for about a hundred years.

But even a galaxy containing these myriads of stars is but a minute fraction of the known universe; it is estimated that the number of galaxies is also 10^{11}, so that there are as many galaxies in the universe as stars in our Galaxy (see Section 3.14).

4.8 Starlight

If for a moment we can imagine all stars to be at an equal distance from the Earth, then their brightness would vary over a range which we can represent as "from lighthouse to glowworm." In this scale our sun would be represented by a candle.

The star S. Doradus in the constellation "Goldfish," one of the stars in the Greater Magellanic Cloud, corresponds to the lighthouse, with 1 million times the sun's brightness. If this star replaced our sun, within a few moments all our forests would be ablaze and our oceans dried up.

On the other hand with the star Wolf 357—corresponding in brightness with the glowworm in our light scale—in place of the sun, at only 1/5000 of the sun's brightness, the temperature at the Earth's surface would drop to near absolute zero, and the air would become a very cold liquid.

4.9 10^{20}-Electron Volt Protons [21]

Part of the protons reaching the Earth's atmosphere from outer space (see Section 8.12) are at a level of 10^{20} electron volts (eV).

What this means in terms of more familiar units of energy may be illustrated by the statement that if the energy of one of these protons were to be absorbed in the head of a pin, 0.5 mm in diameter, the temperature would rise to 65,000°C; otherwise expressed, this energy from one proton would be enough to melt the heads of 40 pins at a temperature of 1500°C.

The protons of this high energy contained in a sphere of only 10^{-3} cm across (much smaller than the pinhead) would supply enough energy to melt a mass of iron as large as the earth.

4.10 Quasars [47]

A few years ago certain starlike sources of an enormous output of radio energy, receding at velocities comparable to those of the most distant galaxies, were discovered. More than 120 of such quasi-stellar objects or "quasars" are now known.

One of them (3C 273B) produces even 10^{56} ergs of energy per year. This is a thousand million million (10^{15}) times the power liberated by the sun. If the lifetime of such a quasar is only 10,000 years, this means that its total emitted energy is equivalent to a million times the rest mass (in terms of the Einstein formula $E = mc^2$) of the sun. Other quasars emit in the infra-red range at a brightness implying that a mass corresponding to a whole galaxy is compressed in a volume only a few light-years across.

Let us be thankful that this quasar is 10 light-years remote and not 8 light-minutes, as our sun!

4.11 Comets [22]

Comets are believed to be a loose assembly of small metallic particles flying through space, as does a swarm of bees through the air, with icelike lumps of frozen "gases," for example solid ammonia, interspersed among them.

They are seen to be relatively transparent and do not dim the light of stars seen through them to any noticeable extent.

Comets are known to have only a very small perturbing effect on the orbit of a planet which they may pass closely or even envelop, and from this astronomers conclude that their mass is small for a heavenly body, at most a few thousand tons.

This few thousand tons, while a big load on earth, is a very small mass for a body of which the head is as large as the distance from the Earth to the moon, and which, with its tail, may reach from the Earth to the sun. It is hardly surprising that the comet's average density is so low that it would be termed a "good vacuum" in an earthly laboratory (see Section 10.1).

4.12 The Thrifty Sun, the Lavish Bacterium

Although the very high surface temperature of the sun, 6000°C, assists the radiation of energy (proportionally to the temperature), nevertheless the rate of heat release calculated per unit volume of the sun is relatively low.

As a consequence of its large volume (1.4×10^{18} km^3) its surface is relatively small (see Section 8.18A), and because the sun is in a vacuum, heat is given off only by radiation and none by convection. In fact per cubic centimeter of the sun's volume the total heat release is 2×10^{-4} calories per hour.

The situation is entirely different in the case of a bacterium with its tiny volume, 10^{-15} liter. This creature has, moreover, plenty of opportunity to dispose of energy by convection through its usually aqueous surroundings, whereas by its movements mechanical energy is produced as well as heat. Taken together, these amount to 400 calories per hour per cubic centimeter volume of bacteria, thus 2 million times the sun's rate of heat release based on volume.

And now what of a man? On the volume scale he is nearer to a bacterium and this determines the outcome. With 2 calories per hour per cubic centimeter of body volume, the man does 10,000 times better than the sun's rate, but 100 times less than that of a bacterium.

We are fortunate that the sun is thus sparing with its release of energy. Had this release been as fast as that of a man, then the Earth would now be too cold for us to live on.

4.13 The Solar Wind

The solar wind is the name given to streams of particles emitted from the sun and comprised mainly of photons and electrons. Even in periods of "quiet sun"—that is, few or no sunspots—this wind has a velocity 10,000 times that of a "100 km per hour" hurricane.

Why are we not blown off the Earth, and why is the only noticeable effect of this racing wind a certain degree of electromagnetic disturbance in the atmosphere? For all its speed, as material it is very tenuous indeed; the density of the solar wind is only something like 10 protons per cubic centimeter—a good laboratory vacuum, like that of a comet (see Section 4.11).

4.14 Taking the Sun's Temperature

● The temperature in the interior of a star is believed to reach 20 million °C at the center.

Supposing that our thermometer was graduated at 1 mm per 1°C, then to register the sun's surface temperature, 6000°C, the mercury would rise to 6 meters. But to find out the sun's temperature near the center with such an instrument, it would need to be as long as the distance across (Greater) London from north to south, about 20 km.

● It is reassuring to know that even if the primary chemical reactions in the interior of the sun were to cease, far from plunging headlong into thicker underwear, we should blithely continue to wallow in our sun baths for a period of time twice as long as the ancient Sphinxes have been contemplating the Egyptian desert (2×5000 years). For, in their passage to the surface of the sun, the primary radiations react upon surrounding atoms and electrons, thus evoking secondary radiations which follow increasingly haphazard paths and require on average 10,000 years to reach the photosphere.

Assuming that no *gradual* extinction of the sun's energy took place, we could thus expect to bronze our skins until the very last dose of secondary radiations had reached the Sun's surface, but we would then be wise to have mitts and muffs to hand, because in only 8 minutes the climate would become distinctly and permanently chilly!

4.15 The Sun's Flaming Surface

The photosphere, a 1000-km thick layer at the solar surface where most of the sun's rays originate, has a grained appearance when photographed from the Earth. The grains are thought to be the crests of gas fountains which spring from the interior of the sun.

The Vatican astronomer Secchi has compared the flaming surface of the sun with a rice pudding, the individual grains of which can be as large as the Earth.

4.16 Colonizing the Moon

Spores of bacteria are not killed by immersion in either liquid helium (−269 °C) or boiling water (see Section 8.15). Nevertheless conditions on the immediate surface of the moon are so hard—there is lack of oxygen and water, but there is heavy radiation—that this desert is considered beyond the colonizing capabilites of the most versatile microorganisms. But a few inches down under the dust conditions may be very different.

Colonization might occur if even a single microorganism were introduced.

Colonization of the moon might occur even if only a single microorganism were introduced. Assuming one division into two daughter cells every month—slow for a microorganism—the intruder would become widespread within 2.5 years, providing it were suitably dispersed by wind all over the planet. The following time schedule for this sort of colonization has been suggested [12]:

- after $2\frac{1}{2}$ years contamination becomes widespread
- after 4 years one organism per square meter of the surface
- after 5 years about a ton of organisms in total
- after 8 years marked effects on soil and surface chemistry

4.17 Hibernation on the Moon

Someone living on the moon, anywhere except very near its poles, would experience a difference between "day" and "night" temperatures of about 190°C, that is from 130°C by day down to −60°C by night. This enormous change would take place, upwards or downwards, once each 2 (earthly) weeks, for the moon's "day," one complete rotation about its axis, occupies about 4 earthly weeks.

This change in temperature is about $2\frac{1}{2}$ times as much as the change experienced by someone who by some means could pass from the Earth's South Pole to the equator in one step.

Hence H. G. Wells' comment: If there is any sort of life on the moon, the creatures must hibernate through the lunar night and start afresh each lunar day.

4.18 "Instantaneous" Hardening of the Surface of the Moon

Successful "soft" landings of experimental space vehicles on the moon have shown that the surface is not covered by a layer of dustlike material, as was supposed in many quarters, but is a hard rocky crust, whatever may lie beneath it.

Research has indicated the possibility that this hard crust is an effect of radiation from the sun. On the laboratory scale, it has actually been demonstrated that sand became rocklike in consistency after a "dose" of 6×10^{17} neutrons. On the basis of this sort of result, it is estimated that a layer of sand or dust 1 cm thick on the surface of the moon would, under radiation from the sun, change to a rocklike substance in

10,000 years—in cosmological terms, an almost instantaneous hardening.

4.19 Some Astronomical Bills

● *The Aerospace Industry*

In 1963 the newly born aerospace industry in the United States surpassed in many respects the country's mighty automobile industry. Including the building of airplanes, military missiles, and hardware for moon flights, the total annual payroll of nearly 10,000 million dollars—representing a contribution of $50 by the average American—was parceled out among 1.3 million employees, a group exceeding in number the population of Baltimore, Maryland, by about 25%. By contrast the automobile industry employed some 25% less people than the Baltimore population (800,000 employees).

● *Circumterrestrial Flights*

The circumterrestrial flight of the kind undertaken by John Glenn in 1962 cost in the region of 3 million dollars, a mere 1.5 cents for each American.

● *Moon Photographs*

The journey of Ranger VII in 1964, during which American space scientists were able to photograph the surface of the moon from a distance about equal to the height of the Eiffel Tower—300 meters (about 1000 feet)—cost $260,000,000.

The journal *News and World Report* published some other possible uses for this sum of money:

give $4,000 to every American family

or

finance the U.S. program of help for underdeveloped countries for 20 years

or

construct 33,000 km (about 5500 miles) of motor highways

or

discharge all mortgages on farms in the U.S.A.

or

finance the American armaments program for 16 months

• *Launching the Space Laboratory*

Space scientists in the United States are in hopes that around 1970 they will have at their disposal a "space laboratory," orbiting the earth, with a crew of 36 and supplies of food and oxygen to last for a year.

This laboratory is to be launched by the immensely powerful "Saturn V" rocket; some vital statistics of this giant follow:

Height:	About that of the spires of Notre Dame cathedral in Paris (106 meters, about 350 feet)
Weight:	One-third that of the Eiffel Tower (2700 tons)
Payload:	That is, the laboratory to be "satellitized": About the weight of the obelisk in the Place de la Concorde (100 tons)
Cost:	$30,000 million, one and a half times the French national budget. Or in other terms: An American family with four children will contribute $900—the cost of a new small car—toward the expenses of this undertaking.

• The Saturn V rocket launched from Cape Kennedy blasted off with as much power as 3000 of the biggest diesel locomotives. The cavernous tanks that fueled the engine during its $2\frac{1}{2}$-minute spurt contained enough liquid to fill 50 railroad car tanks (500,000 gallons).

• *Buildings*

The vertical Assembly Building in which moon rockets are assembled is the world's largest building. It could contain the largest Egyptian pyramid: the United Nations building in New York could be passed through the entrance gate.

4.20 The Voyage of Mariner IV

When after traveling a good 400 million miles (600 million km) the fading signals of the Mariner IV launched in 1964, dropped below the threshold of sensitivity of the master receiver at the Goldstone tracking station (California), the signal had a strength of only 10^{-18} watt, two orders of magnitude below the human threshold of hearing.

4.21 Odd Bits

• The average kinetic energy of the galaxies' rotations is the same as the average kinetic energy of their translatory motions. This, a consequence of the general law of statistical mechanics, also holds for the rotational and translatory motions of molecules in a gas. (G. Gamow.)

● The Hubble radius is the distance at which in the theory of the expanding universe the expanding rate is equal to the velocity of light. A sphere with this radius comprises as many galaxies as this Hubble radius comprises light-years (10^{10}).

● The smallest "things" that we can see on the sun—by photographic means, such as sunspots—are about as large as France, with a diameter of 1000 km (600 miles).

● The mass of one man stands in the same ratio to the mass of one atom of hydrogen as the mass of the sun to the mass of the man with whom we began.

● The surface area of the moon is greater than that of Africa, but less than that of North, South, and Central America combined.

Chapter 5

Our Planet

To build for the future means leaving a great deal unbuilt, for nobody knows the demand of the future.
(F. van Klingeren)

5.1 From Adam to a "Doomsday"—A Surfeit of Calories [4]

Scientists, while agreeing that the provision of food for Earth's growing population is not likely to be the limiting factor for that growth, maintain that the "squeeze-to-death" limit is altogether too modest.

To accommodate a really striking increase, or rather multiplication, of the Earth's inhabitants, they propose the use of satellites mounted with reflectors so as to make much more of the sun's energy available to man, and they even propose the installation of filters to limit this energy to the useful frequencies. With synthetic food production—agriculture, animal husbandry, and fishing being long outdated—it is estimated that a population density of 5 per square meter might be supported.

Here we are obviously in a squeeze-to-death situation, but we may

We are obviously in a squeeze-to-death situation.

assume that building technology will have kept pace with other developments, and that tower blocks reaching up even to the stratosphere will become commonplace; of course the upper stories would have to withstand near-vacuum outside and be pressurized like an aircraft inside. Moreover, such constructions need not be confined to the Earth's land surface; for those who might not enjoy height, there is the interesting possibility of a flat, 60 stories *down*, somewhere in the seas or oceans.

However, to maintain the energy needs of this sort of population, so much radiation would have to be picked up by satellite mirrors, and other such devices, that in the course of not many years the polar regions would have become as warm as the tropics now are, and if no solution was found the earth's oceans would gradually rise to the boil.

Even this problem can be solved, however, if means could be found of converting the low-grade waste heat from the earth's activites back to high-grade energy. This can be achieved to some extent by heat pumps, mechanisms working on the principle of the household refrigerator (that is, abstracting warmth from the contents of the refrigerator and channeling the heat out of the cabinet, usually at the back). These pumps would let the Earth's inhabitants dispose of surplus unusable heat, sending it back into outer space.

The ultimate limit under these conditions seems to be a population of about 10^{16}—10,000 million million—and to go beyond this would mean a gradual warming up of an Earth, already quite remarkable in its characteristics, to a temperature intolerable by man as he now is. But, even so, the population density would be about 120 per square meter, say 8 tons or a half-load for a railway wagon. Anything like a natural landscape would have become extremely rare, and one's afternoon excursion would probably be up and down, rather than round about, a region dotted with towers reaching above the clouds. But what about the idea of having a chance to meet a General de Gaulle, because there would then be 10 million such persons [64].

5.2 Due to the Earth's Rotation

● *The Earth as an Electric Generator* [23]

The rotation of a celestial body in which there exists a magnetic field results in the development of an electrical potential, commonly known as homopolar or unipolar induction.

The Earth, revolving once every 24 hours, carries with it a magnetic field of about 0.1 gauss, so we may see the Earth as a reservoir of energy of something like 7×10^{22} kilowatt-hours. If we could harness this—accepting that the Earth would gradually cease to rotate about its axis—it would be power sufficient to supply energy at the level of our present needs for half the time which has elapsed since the Earth's surface became solid.

Indeed, if we see the Earth as the rotor of a natural generator with the poles "+" and the equator "−," we may calculate that a potential difference of 100,000 volts exists between equator and poles.

● *Polarifugal Forces* [12]

As a result of the rotation of the Earth, centrifugal force acts on the land masses tending, particularly in the middle latitudes, to shift them toward the equator. Near the poles the forces are however small, and near the equator the force is of course "upward"—away from the center of the Earth—rather than sideways.

But, at the latitude of Manhattan Island, it is estimated that the Earth's rotation produces a "polarifugal" force of 50 kg on a section of ground 1 meter square and 10 meters deep; so the total force acting to displace New York City's business center—some 80 km^2, about 20,000 acres—is of the same magnitude as the pull exerted by 5000 transatlantic liners of the *Queen Elizabeth* type tied up along the embankment of Battery Park but attempting to steam southward at full power.

5.3 The Sea as a Heat Accumulator

To freeze all the water on the Earth from an assumed average temperature of 10°C, a quantity of heat would have to be abstracted from it about equal to the heat received as radiant energy from the sun in the last 100 years, that is, 10^{23} kilogram calories.

So, during the onset of a future Ice Age, the heat needs of two or three human generations could be provided from this source at least quantitatively, though (directly) one gets the impression of a chilly business.

Should our Earth's climate change for some reason to something much warmer, it is a comforting thought that for all the water in the seas and oceans to be boiled away, the heat needed would be as much as that received as radiant energy from the sun since around the year 1300 A.D. (that is, when the Crusades finished), 7×10^{23} kilogram calories.

However, if that sort of change in climate should ever come about, we should need heat insulation on our houses to an extreme degree—and a very good cooling system inside—to protect ourselves from being "boiled away" too.

5.4 The "Biosphere"

The diameter of the Earth is about 13,000 km (8000 miles), but the "biosphere," that is, the layer or zones where life is found, is less than 20 km from top to bottom, in fact about 18.5 km ($11\frac{1}{2}$ miles, or 60,000 feet) on latitudes of moderate temperature.

On a terrestrial globe 1 meter in diameter (about 3 feet) this thickness would be represented by 1.5 mm, and the highest mountains and deepest troughs—the extreme range of life in contact with the Earth—would be layers just less than $\frac{3}{4}$ mm thick, or scratches just more than $\frac{3}{4}$ mm deep. But they would represent heights up to nearly 9000 meters (29,000 feet) and depths of nearly 10,000 meters (32,000 feet).

5.5 Attraction by Celestial Bodies

A physicist, interested in the forces of attraction between bodies, consulted his *Book of Handy Tables* which happened to weigh $\frac{1}{2}$ kg, about 1 lb. He calculated that its weight on the sun's surface would be about 12 kg, or 26 lb, the weight on Earth of an old family Bible, but for a moon explorer, the weight there would be only about $\frac{1}{10}$ kg, or $\frac{1}{6}$ lb, the weight on Earth of a pocket diary.

5.6 Our Threefold Rotation

A man standing on the Earth's surface, has, relative to the nearer galaxies, a remarkably complex motion.

To begin with, from the Earth's rotation about its own axis, he is moving with a circumferential speed—if he happens to be at the equator—of 1600 km (1000 miles) per hour, about one-half of the speed planned for coming commercial supersonic passenger aircraft.

Second, the Earth's annual rotation around the sun takes him through space at about 110,000 km (70,000 miles) per hour, that is at a "circumferential" speed of 35 times that of the planned aircraft.

Third, the sun and the solar system are in slow rotation about a point within our Galaxy, and although the sun and its retinue of planets make, together with the Galaxy as a whole, only one circuit in 200 million years, this adds to our friend's motion a speed of 1.6 million km (1 million miles) per hour relative to the nearer galaxies; this is 500 times the planned aircraft speed.

Now our friend takes a train!

When we say that the train has moved 10 km (6 miles) nearer to New York Central Station, we mean that it has moved a distance of 10 km along a pair of rails along which it is running to the station. In the interval of time, say in 5 minutes, the Earth—carrying this pair of rails with it—has carried the train [71]:

- 130 km to the east by its daily rotation around its axis;
- 9000 km in its earthly orbit around the sun (the distance from New York to Cairo);
- 110,000 km farther away from a distant nebula, due to the motion of our Galaxy as a whole (nearly 3 times around the Earth).

5.7 A Circle or an Ellipse? [24]

Jimmy, when told to draw a diagram of the Earth's orbit about the sun, stepped to the blackboard and drew a circle.

"But Mr. Armistead, sir . . . you told us it was an ellipse!" exclaimed little Mary.

"You are right, Mary, but as a matter of fact Jimmy's circle is a tolerable representation of the Earth's elliptical orbit, its variations from a truly circular path being contained, on his chosen scale, well within the thickness of the chalk line."

5.8 Errors in Our Chronology [25]

A Small Error

Our present Western calendar dates from the Papal Bull "Inter Gravissimas" in which Pope Gregory XIII decreed the omission of 10 days, and the reduction in the number of leap years—every fourth year as established by Julius Caesar—by limiting the centesimal leap years to those exactly divisible by 400, so that the centesimal years 1700, 1800, and 1900 were to be ordinary years.

This calendar was to commence on 15 October 1582, and Gregory's ingenious adjustment resulted in a mean "Gregorian" year comprising 365.2425 days. This is so near to the true solar year, 365.2422 days that in the 386 years which have elapsed since his decree the calendar error is only 3 hours.

● *A Very Small Error*

The Russian scientist Kotelnikof, investigating the (very) gradual slowdown in the Earth's rotation consequent on tidal action, found that between 1963 and 1965 the average length of the Earth's day increased by 1.6 milliseconds.

Suppose a newborn baby were to be given the handsome present of two clocks, a grandfather clock from the days of the czars, regulated correctly relative to the length of the day 50 years ago, and something more modern, regulated correctly relative to today's length of the day. When the time came for our baby to retire after a useful lifetime's work at age say 65, his old grandfather clock would be 37 seconds fast compared with the other more modern "birthday" clock.

● *A Still Smaller Error?*

Very accurate measurements have shown that an eruption on the sun which took place in 1959 has caused the earth's rotation to be slowed down by 1 millisecond per 24 hours. Who calculates how much such erruptions, by their cumulative effect, will make the Earth's days little by little longer?

Errors between Earthbound Clocks: Some Effects of Place on Time [70]

Many will recall how, in the early days of pendulum clocks, differences were found in the performances of the same clock in the tropics and in Europe, and also at the top and the bottom of a high mountain. These quite substantial discrepancies were traced to variations in "*g*," the acceleration caused by gravity, consequent on centrifugal force and on differing distances from the Earth's center.

Time measurement has now become so accurate that differences predicted by Einstein's relativity theory, for observers at different points on the Earth, can at least be discussed although the practical recording of the differences may be difficult.

In the first place, the relativity theory predicts differences caused by varying gravitational potential, as found when ascending a high mountain. It has been calculated that for Mount Everest, 8888 meters, or 29,000 feet high, allowing also for the effect of increased centrifugal force, the difference between the readings of two mechanically perfect timepieces would, after a year, be 31 microseconds. By what means of instantaneous communication the observer on the summit could make contact with his friend down a shaft or in a tunnel at the bottom of Mount Everest, we leave others to work out, but timepieces are being developed stable to within 1 part in 50 million million, which is quite accurate enough for the purpose.

Further, the theory predicts differences of a similar magnitude due to the different speeds of movement of different points on the Earth's surface. As between one of the poles and the equator, the difference in the course of a year should amount to 38 microseconds. Measuring this again confronts us with the problem of instantaneous communication, for an exchange by radio about the progress of the clocks as they approached their midnight winning-post would occupy 30,000 microseconds for the transmission of any sign by radio round one-quarter of the globe.

5.9 Earthquakes

Although 1 million earthquakes are recorded each year on the world's seismographs, only about 5000 of the tremors or shocks are felt by man as he goes about his daily life, and of these, in an average year only 100 cause actual damage and possible loss of life.

So the great earthquake, recorded in Alaska in 1899, one of the most severe known, was indeed exceptional; it has been estimated that the energy released in and by it was—in ballistic terms—equal to the explosion of 20 million tons of TNT. To equal this we should need 1000 "Hiroshima" bombs—or 7 sharp thunderstorms (see Sections 12.5 and 12.6).

5.10 Stardust

On a conservative estimate, about 100 tons of meteorites and other solid matter from outer space reach the Earth's surface every 24 hours, that is, enough to load 5 large railway wagons. All this material that has fallen, since the Earth's formation, spread over its land and sea area amounts to a layer only a few centimeters deep, negligible when compared to the material removed by wind and water erosion.

Though the daily shower includes millions of visible particles, larger at any rate than the head of a pin, the danger of really large objects penetrating the Earth's atmosphere right down to its surface without being vaporized by the heat of atmospheric friction is fortunately very small: perhaps 10 of them weighing 10 kg each, per day.

Nevertheless, there is a definite hazard: a 60-ton meteorite has been found in Greenland, and one of 150 tons weight in South Africa, and it is supposed that a very much larger specimen fell in the state of Arizona, to form the well-known crater which is about 1200 meters (1300 yards) across.

This Arizona meteorite is estimated to have weighed 2,000,000 tons, and to have struck the earth's surface at—at least—"escape velocity" (40,000 km per hour). This velocity implies that its kinetic energy must have been of the order of that of a 30-megaton TNT bomb, and so 50% greater than the explosive power of the Hiroshima atomic bomb.

The possibility of a giant meteorite's striking the earth once more is by no means only imaginary; the minor planet or asteroid Icarus approached the earth within 6 million kilometers—20 times the moon's distance—in 1949, and this near approach recurs every 2 years. Icarus is thought to weight 30,000 times as much as the Arizona meteorite, and its fall on earth would be the equivalent of 40,000 "Hiroshima" bombs.

If it became known that such a collision was a possibility, possible means of averting the catastrophe would be to fire a hydrogen bomb at the approaching body with the object of reducing it to powder, or on a more modest scale, firing some fairly heavy projectile so as to change its orbit sufficiently to make it bypass the earth at least "this time."

While thinking about these giant objects we must not forget minutiae, that is the neutrinos of which 10^{10} are estimated to strike each square centimeter of the earth's surface every second of which, however, only 10 or so are absorbed (see Section 2.8), nor must we forget that protons also reach us from outer space (see Sections 4.9 and 8.12).

5.11 A Worthless Shower of Gold

Although per second there falls on every square meter of the Earth 25,000 neutrons, of which 1 gram weight can be compared in value to the cost of building a 15,000-ton freighter (by virtue of its costly production), there is no profit in trying to exploit this "golden rain".

For, the value of this fall in 1 second over the Earth's whole surface is only $25.000, consequent on the trifling weight of one neutron, that is 1.7×10^{-14} gram. How unlucky it is that meteorites do not have the same value, weight-for-weight, as neutrons, since for every gram of neutrons which reach the Earth, about one thousand 10-ton wagon-loads of meteoric material comes down (see Section 5.10).

5.12 Salt, Salt, Salt

A visitor to the Great Salt Lake in Utah, perhaps dazzled by reflections from the glittering salt beds along its shores, turned to mental arithmetic . . .

He thought of all the salt in the sea; if this were to be collected and spread in granular form over the Earth's land area, the layer of salt would about cover the Washington Monument, or Saint Paul's cathedral, or the well-known spires of Cologne cathedral, being about 165 meters (about 500 feet) deep.

Salt accumulation in the sea—in which its concentration is now about 3.5%—is going on at an ever-increasing rate, because, as time passes, material which was originally solid rock is constantly being broken down to smaller pieces and even to sand, so that any salt present is dissolved more and more rapidly. Our mental arithmetician estimated that if the present results of this "weathering" had materialized immediately at the beginning of the Earth's history, the present concentration of salt in sea water would have been reached after 300 million years (3×10^8), thus long before there was any question of life on the Earth's surface.

Now for the future: because of the steadily quickening transfer of salt from land to sea, it seems certain that within a period equal to that which has elapsed since the Earth solidified, that is 3500 million years (3.5×10^9), the concentration of salt in the sea will have reached saturation point, 36%. Our successors will find—at that far-distant time—that all the seas and oceans have turned into something like the Dead Sea or the Great Salt Lake.

À Propos

The Dead Sea is estimated to contain about 20 tons of valuable salts for every man, woman, and child on the Earth. The Great Salt Lake, though greater in size, is less concentrated, and could yield perhaps half this quantity.

5.13 The Disappearing Land

Roughly two-thirds of the surface of the Earth is covered by water. The area of the Pacific Ocean alone is actually 25% larger than that of all the land surfaces of the world taken together.

One of the reasons why the land and the sea level approach each other more and more is because land is washed into the sea by rivers.

Taking the average height of the United States above sea level as 800 meters (2600 feet) and putting the rate of erosion at 6.3 cm (2.5 inches) in 1000 years, the volume of the United States above sea level will be washed to the ocean within 12 million years: this is only something like one-tenth of the period of time since mammals appeared on our planet.

Another cause of the gradual disappearance of land is that for a good half-century the chimneys of our factories and of an increasing number of our houses have been pouring carbon dioxide into the atmosphere, so that the proportion has increased appreciably from its nineteenth-century value of 0.029%. In fact, the increase since 1900 is estimated at 12%, that is between 0.003 and 0.004% of carbon dioxide in the atmosphere.

Since carbon dioxide in the atmosphere slows down the heat loss from the Earth by radiation, the average temperature at the surface of our planet has risen in the same period by 1.1°C. If things continue as they are for 1000 years, then tropical or subtropical climates would prevail over the whole surface of the Earth; the polar ice caps would melt, and all land now within a few meters of sea level would be engulfed.

Western Europe is already subsiding at the rate of 2.5 cm (1 inch) every 10 years, and, if this continues, the top of the Eiffel Tower will become a small island in the Atlantic in roughly 200,000 years.

5.14 Daily Intoxications [15]

● *Intoxicating the Air*

As we drive our car into town, for every liter (about one-quarter of a gallon) of gasoline used, about 15 cubic meters (500 cubic feet) of gas issues from the exhaust pipe; luckily a considerable proportion of this is nitrogen. However, as we drive along with our speed carefully limited to 50 km (or 30 miles) per hour, we leave a trail of the poisonous carbon monoxide; half a meter from the exhaust outlet there is about 1 volume in 30 present in the air. Although at 10 meters (or 11 yards), this has fallen to 1 volume in 10,000, this is still the presently permitted limit in many countries, or in other words, the concentration which it is considered can be safely breathed through a working day of 8 hours, without bodily injury.

One may thus conclude that in town or city traffic every car has behind it a (probably) pear-shaped mass of gas some 10 meters (11 yards) long, within which the permissible "safe" concentration of carbon monoxide is constantly exceeded.

● *Exhaust Gases and Cigarette Smoke*

The safe limit of carbon monoxide in air being 1 volume in 10,000 volumes, somebody breathing the usual amount of air, about 40,000 cm^3 per hour, would take in 4 cm^3 of carbon monoxide if in an atmosphere at the safety limit. Dense city traffic with 2000 cars passing per hour produces a concentration of 1 volume in 40,000, well clear of the safety limit; but at a busy crossing where there is much stopping and waiting the concentration of carbon monoxide may rise to 1 volume in only 5000, well in excess of the "safe" figure.

Even so, one is in less danger from carbon monoxide poisoning at a busy city crossroad than when inhaling cigarette smoke. Taking 50 cm^3 as one "inhalation," this brings with it 1 cm^3 of carbon monoxide. Thus smoking only one cigarette, and this with only 10 inhalations, is the equivalent, in carbon monoxide terms, of:

- $1\frac{1}{4}$ hours at a busy city crossing
- $2\frac{1}{2}$ hours in an atmosphere at the limit of safety
- 10 hours in dense, but moving, traffic in a town or city

In this connection it may be noted that world production of cigarettes is currently quite sufficient to provide every living person with two cigarettes a day (2,500,000 million per year). If we all smoked this quota, with inhaling, we should each take in the same amount of carbon monoxide as would result from standing in a busy city street for 20 hours, or from 5 hours of exposure in an atmosphere at the safety limit.

On the one hand this would suggest the merit of residence at high or remote locations to avoid the exhaust-gas peril. But on the other hand this must not be carried too far (or too high!), because residence at say 2000 meters (6500 feet) above sea level inactivates as much hemoglobin as the smoking of 10 cigarettes a day—however, in this case without inhaling.

5.15 Nutrient Cycling in the Air

The carbon dioxide in the atmosphere, weighing 15,000 times as much as the total world population (altogether 25×10^{14} kg), would be exhausted in a year or so by green plants if the atmosphere were not continuously recharged with CO_2 generated by respiration and by fires.

5.16 Urbanization [26]

Taking a country's people to be "urbanized" when about one-third are living in towns and cities of population 100,000 or more, we may follow its progress over the years and in different countries.

Before the Industrial Revolution, about 1800, urbanization was barely perceptible even in Europe—

in 1600 1.6% of Europeans lived in towns of 100,000 (or more) population

in 1700 this percentage was 1.9

in 1800 this percentage was 2.2

The transformation started in England; by 1801 one-tenth of this country's people were living in towns or cities of 100,000 (or more) population, and by 1840 the country could be considered to be "semi-urbanized" with some 20% of its population in agglomerations in the "100,000+" size.

. . . semiurbanized in 1840.

In general, the later a country became industrialized, the faster was the process of urbanization, as appears from the growth of the urban proportion in nations older and younger in terms of industrial development.

The transformation from a population with 10% of its members in towns and cities of 100,000 or more into a society with 30% of its people in these took:

in England . . . . the lifetime of an old man (80 years)

in the United States . . . a period equal to the usual age at retirement (65 years)
in Germany . . . . a "middle-age" lifetime (50 years)
in Japan . . . . the time it usually takes to become "well established" (35 years)
in Australia . . . . what we call "not long married" (25 years)

The world as a whole has now attained the level reached by England in 1840, i.e., with 20% of the population living in units in the 100,000+ size, which we have called "semiurbanized." If the increase in urbanization continues at its present rate, by the year 1990 more than half the world's population will be living in the "100,000+" agglomerations, a situation which we may call "heavily urbanized."

There now exist 1375 centers of this "100,000+" type; perhaps to our surprise, we find about half of these "large" concentrations in the Eastern countries. The same 50:50 distribution between East and West holds for the 36 "larger" centers, with 200,000 or more. The total number of people living in these 200,000+ units is comparable to the number of "whites" in the United States (150 million).

At the top of the list we have two mammoth agglomerations each with a population like that of a whole nation or state. New York City's

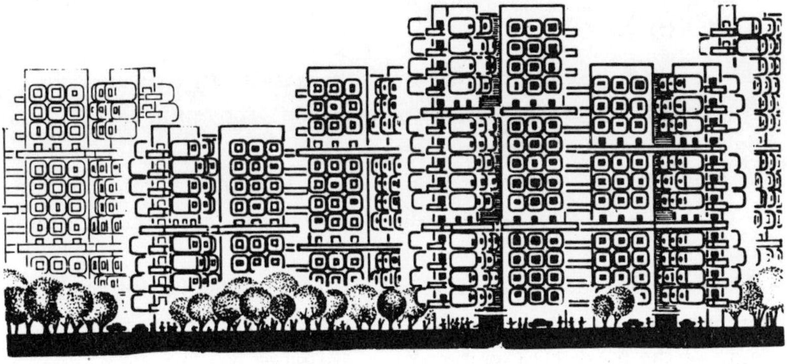

. . . present mammoth agglomerations have a population like that of a nation.

14 million citizens outnumber the population of The Netherlands, or that of the state of Texas, and Tokyo's 10 million about equal the population of Belgium, or the state of Illinois.

A final note—by 1960, as much as 50% of the population of the

United States was concentrated in about 200 heavily urbanized areas, which together occupied only 7% of the nation's land.

5.17 Hunger

It is estimated that the total worldwide loss caused by insects and other pests amounts to about 35% of important crops such as grain, potatoes, and sugar beet, and, at the same time, the undernourished are estimated to number about 30% of all mankind, among whom 10% are really hungry.

In spite of this correspondence in numbers, the problem of feeding the world's population adequately would not be solved by the elimination of various pests, and so of the crop losses which they cause. Why not? Because every morning the world's breakfast table needs as many new places—160,000—as the population of a fair-sized town or city, say Norwich, England or New Haven, Connecticut.

It should not be as difficult as sometimes appears to give useful help to the "hungry half" of the world population. The Netherlands has in most years a surplus of butter, which, if put up in the usual $\frac{1}{2}$-lb packets, could form a continuous row from Amsterdam to New Delhi.

5.18 Odd Bits

- The pull of gravity becomes less as we are further away from the earth. In an aircraft, flying 15 miles (24 km) above the earth, a man weighs about 1 lb (450 grams) less.
- A ton—1000 kg—of stone at the top of the Empire State building in New York City weighs 100 grams less than its weight on the ground floor.
- The greatest load imposed on the earth's crust is the 3×10^7 km^3 of ice overlying Antarctica, the South Pole continent. This is 7 times the weight of the Mediterranean.
- Since the epoch of maximum glaciation the volume of ice is estimated to have diminished by one-third (12×10^6 km^3) of its present volume, resulting in a rise in sea level of about 30 meters. Melting of the remaining ice would produce a rise of another 60 meters.
- The tide due to the attraction of the water by the moon is 3 times as large as the tide due to the sun.

● Industrial activity introduces 100 times more lead into the urban environments (e.g., from leaded gasoline) than is leached yearly from soils over the entire earth and added to the oceans.

● A dime (10 cents, in U.S. currency) or similar small coin seen from a distance of 3 km (1.9 miles) just fills the angle "one second of arc" [73].

Chapter 6

The Source of Everything: Energy

Give me energy and I will construct a world out of it.
(freely quoted after I. Kant)

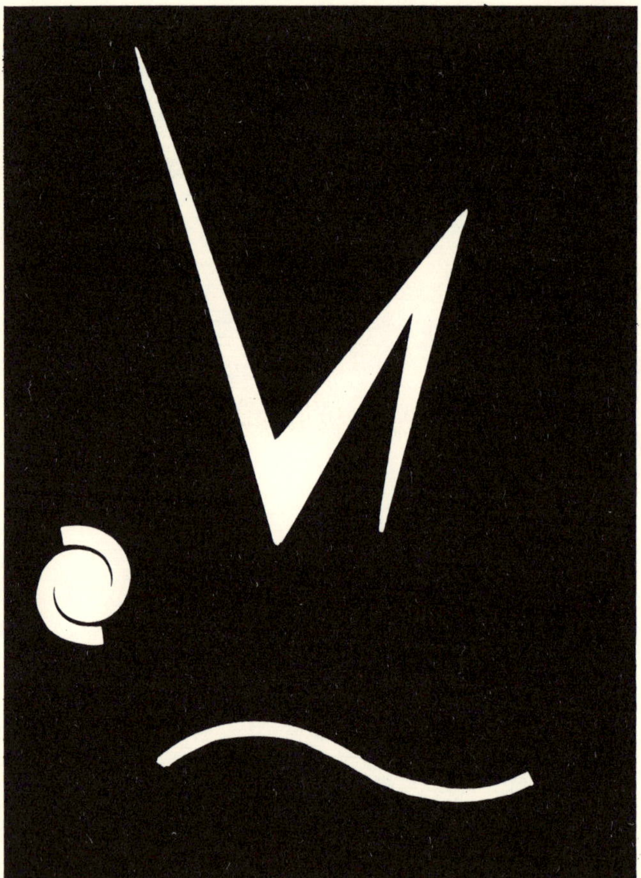

6.1 Energy Requirements and Energy Sources

6.1A Units

The electrician is accustomed to express himself in kilowatt-hours (kWh), the mechanical engineer prefers kilogram-meters (kg m) or foot-pounds (ft lb), whereas the chemist refers to calories. Often very little understanding exists between these experts concerning the "exchange rates" of each other's units, but the following table may be helpful.

Some Basic Energy Data

kWh	kcal	kg m
1	860	3.8×10^5
1.2×10^{-3}	1	430
2.7×10^{-6}	2.3×10^{-3}	1

The purpose of this book would be neglected, however, if such a table were accepted as a final solution in clarification and therefore the following "translations" are given.

● *for the electrician*

1 kWh can light a 100-W lamp for 10 hours (36,000 seconds)
1 kcal can light a 100-W lamp for 45 seconds
1 kg m can light a 100-W lamp for 0.1 second

● *for the chemist*

1 kWh can increase the temperature by 1°C of 860 liters of water
1 kcal can increase the temperature by 1°C of 1 liter of water
1 kg m can increase the temperature by 1°C of 2.5 liters of water

● *for the mechanical engineer*

1 kWh can lift a man (70 kg) 5.4 km
1 kcal can lift a man (70 kg) 6.2 meters
1 kg m can lift a man (70 kg) 1.4 cm

6.1B Man's Energy Consumption

A man's average daily energy consumption as power and food (total 34,000 kcal or 40 kWh per day), the equivalent of 5 kg of coal, corresponds to:

- the permanent use of a 1700-W electric heater, or
- raising him every day to a height of 215 km (135 miles), say 50 times the cruising altitude of modern aircraft.

Of this energy 94% (32,000 kcal) is consumed as power, and only 6% (2000 kcal) as food.

By exchanging the 1700-watt electric heater for seventeen 100-watt lamps we could provide a house with its light with all switches "on," all around the clock; but only one of these lamps would suffice to represent the energy in an average man's daily food intake, and the other sixteen would represent his use of power.

This house may contain seventeen lamps, each 100 watts. Burning them day and night would represent man's average energy consumption. One lamp would represent his daily food intake, the other sixteen express his power requirements.

6.1C Energy Sources [27]

● *Fossil Reserves on Earth*

The present vigorous attempts in exploiting nuclear energy in "atoms for peace programs" are induced by the limited reserves of coal and oil. These are estimated to some 10^{20} kcal, equivalent to something like 1 month of solar energy poured on the Earth's surface. Taking into account the rapidly growing world population and the increasing consumption per capita, these reserves will be exhausted—or at least no longer available at present cost—within a century or so, although we have now used up only 2% of the earth's total fuel supply. The situation can be summarized as follows (in rounded off figures):

- Solar energy intercepted by the Earth (with its atmosphere): 10^{21} kcal per year $= 1.4 \times 10^{11}$ MW

- Present world energy consumption: 10^{17} kcal per year $= 5.5. \times 10^{6}$ MW (doubling every 10 years)
- Fossil reserves perhaps: 10^{20} kcal $= 1.2 \times 10^{14}$ MWh
- Nuclear (uranium, thorium): 10^{28} kcal $= 1.2 \times 10^{22}$ MWh

Consequently it is obvious that energy sources, other than coal and oil, must be exploited if mankind is to survive.

● *Energy from the Sun*

– *The Abundant Sunbath*

The sun continuously transmutes hydrogen into helium in which process so much energy is liberated (3×10^{26} kcal per hour) that, if we were able to completely harness this power, each person on earth could have at his disposal 70,000 times the total power consumption of the United States or half a million times that of a country such as Great Britain or Germany.

The weight of matter converted into energy and lost to the sun per second is equivalent to the freight capacity of about 400 ships each of 10,000 tons.

The weight of matter lost to the sun per second . . . 400 ships each of 10,000 tons.

Owing to its great distance from the sun, the Earth with its surrounding atmosphere intercepts only a tiny fraction ($1{\cdot}5 \times 10^{17}$ kcal per hour) of its supplier's total output. Even so, if we could learn to utilize this still prodigal flood of energy in a perfect way, wide prospects would be opened up.

On the basis of the food and power consumption together of 34,000 kcal per day, solar energy could satisfy the needs of so many people (1.2×10^{14}) that each 2 square meters (22 square feet) of all the land available on earth, could support one man.

Since few of us would wish to find our neighbors so near, we will put this figure in a more attractive way: each of us, if living in a small provincial town of 20,000 people, could avail himself of as much food and power as is now consumed by all his co-citizens together.

In reality, however, we make but little use of the stream of solar energy which bathes our planet. The main reason is that 40% is lost in the atmosphere, although this helps in keeping our globe warm. Since $\frac{1}{3}$ of the Earth's surface is land, we receive where we stand "only" 3×10^{16} kcal per hour, or 20%: $\frac{3}{4}$ of this on the free world.

From the total energy shower on our "atmospherized" planet only 1% or 1.5×10^{15} kcal per hour is directly involved in supporting life (man, animals, plants): a million times the present world's artificial power production. The remainder is either reflected or it is absorbed and later emitted.

The earth intercepts 1.5×10^{17} kcal per hour $= 1.7 \times 10^{14}$ kW $=$ 6.5×10^{19} kg m per hour $= 2.4 \times 10^{14}$ horsepower.

– Harnessing Solar Energy

Up to the present, man has been successful, to some extent, in satisfying his heat and power requirements from the daily stream of solar energy: from the dynamic forces of water or wind and from the sun's radiation. It is estimated, however, that the methods applied in the past will come to provide no more than 15% of man's needs in the course of the next century. But, fortunately, new promising methods for exploiting some of these sources have been discovered, as will appear below.

● *Power from Running Water*

Hydroelectric plants being perhaps the simplest way of producing electricity have a limited capacity for economic exploitation, due to the fact that important differences in levels are required. Altogether perhaps 10% of our power needs can be covered this way in the next century.

● *Energy from the Sea* [28]

On the basis of an average total tidal movement of 60 cm (2 feet) once every 12 hours, an arithmetical summation of the energy stored at "high tide" in all the Earth's seas and oceans indicates that it is enough to supply about 17 million megawatts as a continuous power output. This would cover 50 times man's present use of electricity, but this is of course of only theoretical significance. Exploitation of tidal power is not economically attractive except where the configurations of the coast and of the sea bottom are such that tidal ranges of the order of at least 5 meters (17 feet) occur.

Such cases exist. For example, on the southern coast of Nova Scotia (Canada) high-tide level can be literally the "height of a house"—15 meters (about 50 feet)—above tide level. But the resulting inshore currents, although substantial, are not rapid enough or otherwise suitable for the direct generation of power, and a dammed-off bay is one practical solution of this problem.

The largest tidal hydroelectric generating complex in the world so far is at Chausey, on the north coast of France, with a capacity of 15,000 million kilowatt hours for an average year. This would meet about half the requirements of a country such as Belgium, a rather small proportion of the worldwide demand for electricity.

A larger-scale possibility is to isolate a large inland sea, in which the evaporation loss and the gain through inflow of rivers are markedly different. The obvious case is the Mediterranean, which is estimated to lose more by evaporation than is gained from the inflowing rivers—the system would work more or less in a reverse sense—where 250,000 million kilowatt hours in an average year could be installed. This is twice the electricity requirement of West Germany, indeed 10% of world demand.

However, this is not a short-term scheme; 50 years would have to elapse after the proposed damming of the Straits of Gibraltar before a 50-foot head would become available, the basis for the capacity estimate above.

But the sea could provide more, if we could exploit the following sources [23]:

- thermal gradients in the oceans could cover 5 times man's present annual usage of electricity;
- the small amounts of deuterium in sea water could supply at least 10^{10} years of energy at the present level of requirement.

• *Wind Power*

Despite the apparently simple technology involved, wind power could economically meet only about 0.1% of future power requirements. Even in The Netherlands, where wind power has a long tradition of useful service, windmills would have to be erected along the entire coastline, with their sails capable of catching all the wind up to a height of some 160 meters (500 feet), before the country's present power needs could be satisfied from this source.

● *Subterranean Heat*

Many of us are perhaps inclined to forget the great reserve of energy in the form of subterranean heat, very high temperatures prevailing 1000 miles or so under our feet. The outward flow of heat escaping at the Earth's crust alone is estimated to exceed our power and food requirements by a factor of ten, but due to low temperatures it is admittedly extremely difficult to exploit. Unfortunately(?), only about 1% of the outflow of heat is concentrated in volcanoes where the possibility of utilization is, perhaps, greatest.

● *Mirrors*

Strictly in theory, a system of mirrors arranged to concentrate solar energy could easily supply the power requirements of the Earth's whole population, but this technique, assuming feasibility, is only really useful in the tropics for such operations as domestic cooking. In the more temperate regions the area of the mirrors would have to be up to one-half of that of a typical industrialized country, while vast energy-accumulators would be needed to maintain supplies on cloudy days, not to mention winter conditions.

● *Photoelectricity* [29]

In artificial satellites, use is often made of the effect of solar radiation on photoelectric cells to produce electric energy for the satellites signaling system, which must be capable of communicating with the Earth.

However, the capital investment needed to provide for this conversion of radiation to usable electric energy is far from encouraging to anyone who might think of trying this on a large scale. One would have to reckon with an outlay of about 1 million dollars per kilowatt of installed capacity, a capacity which would be a minute fraction of the capacity of a local generating station, and which would about suffice for rather discreet lighting in one's dining room.

This scale of capital investment is about 8000 times as great as that needed in the local generating station. However, the challenge of generating power from solar radiation, available at no cost, has stimulated activity so much that some people are already thinking in terms of reducing this investment by a factor of 1000.

As to a practical development on these lines, the following gives an idea of present thought:

In a typical hot desert it is possible to absorb solar energy at the rate of 1 kilowatt per square meter (0.84 kW per square yard) during 10 hours of each day. One plan is to electrolyze sea water to yield hydrogen and oxygen, and to transport these gases* through high-pressure (60 atmospheres) mains to points of use, where energy will be regenerated by the combustion of the gases with only water (or water vapor) as product.

This transport of the gases not only is thought to be cheaper than transmission of electrical energy in the usual way, but also has the advantage that the large quantity of gas stored in the high-pressure mains would cover those periods—night or other times—when solar radiation is not available.

For the United States an area of desert only as large as Long Island, that is 4300 square kilometers (1700 square miles) used as a "heat source" would be enough to supply the country's annual electricity needs, 1200 million megawatt-hours; and it would be no problem to find such an area in Nevada or Arizona.

As to water, the annual requirement would be about 400 million cubic meters (88,000 million gallons) which is approximately the volume of Lake Candlewood in the state of Connecticut—but this lake is a long way from the sun-baked deserts of the West.

Britain's requirement of "desert" to satisfy its needs would be met by an area somewhat greater than that of the Isle of Man, which would not be hard to find in Africa, if permission were obtainable, and the water to be electrolyzed would amount to about $\frac{1}{5}$ of the volume of Lake Windermere.

An especially attractive feature of this way of supplying energy is the possibility of fueling road and other vehicles with the compressed gases, which of course burn leaving only water vapor (or condensed water) as a product of combustion—the air pollution problem solved!

● *Plants and Animals*

We have already seen that man's present intake of food per head has an

* Perhaps burning the hydrogen in the air is more attractive.

average value of 2000 kcal per day, which corresponds to about $\frac{1}{3}$ kg of coal or (to express it in a more appetizing fashion) twice this quantity of sugar. Expressed in terms of solar energy, we can say that in order to feed himself, man makes use of only two out of every million calories poured by the sun over the Earth.

The reasons for this extremely inefficient use of the lavish amounts of solar energy available for man's nourishment are threefold:

- relatively very little of the sun's energy reaching the Earth is taken up by plants and animals (much is absorbed by the atmosphere);
- man consumes an extremely small proportion of the Earth's vegetation and animal population;
- both plants and animals perform very poorly in fixing solar energy.

For example, the quantities of carbon photosynthetically fixed from the air by plants corresponds to 1% efficiency in assimilating solar energy [68].

On the latter point, however, there is potential enough for philosophical speculation. Controlled biological photosynthesis, exploiting the mechanism of certain algae (chlorella), could increase efficiency to 10%. This would mean that the acreage now given over to crops, if inundated to shallow depths, could adequately supply ten times the present world population with food.

Since the harvests could be partly eaten and partly burned as fuel, the same acreage could supply even the food requirements of the world far beyond the year 2000, by which time the world's population is expected to be nearly double its present size. Remarkable as it may seem, the volume of water required for this boring one-crop agriculture would actually be less than the amount now needed for the world's productive farmland with its excessive surface evaporation (leaves).

● *Nuclear Energy*

Men have been fortunate to discover just in time how to generate energy by the *fission* of uranium-235. If the age of the Earth were one-tenth greater than it is now estimated to be, uranium-235 (on which we increasingly rely as a reactor fuel) would not be available to us; radioactive decay would by now have led to its extinction.

As it is, present residual amounts of uranium (and thorium) will make it possible for us to have sufficient energy for our needs for many centuries at reasonable cost, provided that we can solve the technical problems of using "fast" neutrons in the so-called "breeder" type of reactor for the generation of nuclear energy. With this equipment, we can expect a fiftyfold increase in the proportion of nuclei which undergo fission; with "slow" neutrons only 0.5% of potentially fissionable nuclei are actually "split."

A power station operated on this basis will consume in a year a weight, in kilograms, of fissile material the same as its power capacity expressed in megawatts.

It is estimated [23], on the basis of the world's presently known reserves, that the amount of fissile materials (uranium and thorium) could yield many times the energy obtainable from all exploitable coal and oil if the poor ores could also be used.

But, the sun demonstrates a still more promising method: transmuting hydrogen into helium by *fusion*. The problem is how to carry out this reaction on Earth under controlled conditions at some 5 or 10 million degrees Celsius. "Magnetic vessels" do not yet belong to standard equipment, but there is time, thanks to fission.

Anyhow the available quantity of hydrogen, the "fuel" in this process, is attractive enough. Taking water only, the 330 million tons for every Earthian contain 36 million tons of hydrogen, with each gram to supply 1.6×10^8 kcal. Lavish amounts!

Both processes, fission and fusion are, however, still far removed from the ideal of *complete* conversion of matter to power. According to the Einstein equation ($E = mc^2$), 1 gram would deliver 2.6×10^{10} kWh: this is 170 times what fission yields. The following example may illustrate the hopes for the future.

An electric power station with a capacity of 250 megawatts (electric), serving a city of a million people (Newark, New Jersey, or Bristol, England) could be run* at full capacity:

- during 1 second . . . . by burning 30 kg (66 lb) of coal
- during 36 days . . . . . by fission of 30 kg of uranium-235
- during 10 months . . . by fusion of 30 kg of hydrogen
- during 140 years . . . . by conversion of 30 kg of matter

* Taking the efficiency of the power plant at 33%
1 kg of coal = 7000 kcal

6.2 The Hourly Tanker

If all the world's petroleum requirements were transported by tanker, then every hour a 200,000-ton tanker would have to be dispatched somewhere in the world to keep supplies up: by 1975 the "hourly" tanker would have to be one of the new 300,000-ton mammoths.

It has been estimated that during that same hour, the Earth's vegetation would abstract enough carbon from the carbon dioxide in the air to form 1000 tons of crude petroleum. Supposing that this crude oil could actually be collected, it would show up as a "coaster" compared with the 300,000-ton mammoth tanker.

Of the 1000 tons of "potential" crude petroleum formed every hour, it seems that 50 tons can be attributed to the breath of man and of animals, with the rest attributable to breakdown of plant material by microbes.

6.3 Efficiency in Power Conversion

Already aware of the fact, stated in Section 6.1B that a handful of coal can theoretically provide enough energy to lift a man to the top of a high mountain, our engineering friend Ernest has meanwhile been calculating what level of efficiency in lifting things we have actually achieved in our technical age. His conclusions are hardly encouraging.

Taking the summit of Mont Blanc (4800 meters = 16,000 feet) as the experimental goal, he finds that 100 grams or so of coal ought to be quite sufficient to transport our man to the top, i.e. at 100% efficiency. It then appears that railway locomotives were invented for the primary purpose of burning fuel as wastefully as possible: in transporting, say, 1000 passengers to the summit, their actual or equivalent consumption of coal would be no less than 50 to 80 times the theoretical optimum:

- diesel or diesel-electric locomotive: 5 kg of coal per passenger;
- steam locomotive: 8 kg of coal per passenger.

Ernest now turns his attention to the gasoline engine and concludes that in the motor car we have certainly found a more economical means of transport. The amount of gasoline necessary to afford four persons transported by car a comfortable view from the top of Mont Blanc turns out to be equivalent to 3500 grams of coal per head—a considerable improvement, of course, but still 35 times the theoretical optimum.

Our intrepid engineer then discards his overalls and sets himself to climb a vertical ladder to the summit. Consuming only the equivalent of 850 grams of coal from the lunch box in his rucksack, he arrives at the top panting no more vigorously than the steam locomotive. His efficiency has thus been $12\frac{1}{2}\%$, and his climb has required hardly one-tenth of the fuel devoured by the locomotive from its tender in bringing a single passenger to the same point.

Ernest, however, is vexed by the thought that he has had to drag the whole weight of his body (70 kg = 150 lb) up the mountain. He therefore constructs a hoist with a counterweight which balances the cage and a fair proportion of his body weight, and he is now able to reach the summit while expending the equivalent of only 400 grams of coal—12 times as efficiently as the train-borne traveler . . . and a good deal more quickly!

Ernest has certainly started something! He and his professional colleagues will be kept busy installing lifts in practically every new tall building. Those who continue to use the stairs presumably need the exercise; they burn sixteen times as many calories doing it the "hard way"!

6.4 Material from Light

Radiation from the sun, when it is directly overhead, has the energy equivalent of about 1 kW per square meter (0.84 kW per square yard), and the relationship formulated by Einstein, $E = mc^2$, enables us to calculate the mass of matter with which this energy would be associated if this radiation were incident on the same square meter.

From this it appears that for this area to have received 1 gram of material associated with the heat and light falling on it, we should have to go back, in time, to when Moses received the Ten Commandments 3000 years ago.

Leaving the Earth's actual surface, and considering the "troposphere" —that level in the atmosphere about 10 km (about 33,000 feet) above the Earth where weather phenomena mostly develop—we find that at any moment this contains solar energy, equivalent to:

- 5 million kWh, that is the energy delivered in 24 hours by a 210-megawatt power station, such as would serve a city of 1 million inhabitants.

or

- the equivalent of 0.2 grams of material (see Section 6.6).

6.5 The Power of Lightning [30]

If he were ready to make use of it, a European motorist could drive his car—assumed to be one of medium size—for an hour on the energy in one average lightning flash, though his American friend, with his "long-drawn-out" automobile could only manage about half an hour's run. Moreover, as in the 1800 thunderstorms estimated to be in progress somewhere in the Earth's atmosphere at any given time, about 100 lightning discharges occur every second, there is enough energy for a drive for every inhabitant of the Earth—an hour in a "compact" or half an hour in a larger car.

It is true that some of these motorists are likely to be victims of this—as it seems—storm of lightning; statistics tell us that the proportion is about 1 in 100,000, or in population terms, 1 out of all the people who live in the old English city of York.

During a really heavy thunderstorm, there may be as many as 1000 flashes which in energy terms would be enough, even for our American friend, for a drive right around the Earth—40,000 km or 25,000 miles. Or, confining ourselves to electricity, the energy is about 100,000 kWh, about half an hour's output from a generating station supplying a center with 1 million inhabitants—as examples, Birmingham, England, or Baltimore, Maryland.

Reverting finally to one flash, those who keep out of traffic jams by not owning a car may be glad to know that the energy content should be enough to supply electricity for a small family for a whole month, if only it could be "saved up" in some way.

All electrical disturbances together going on in the Earth's atmosphere, constitute an amount of energy which, if it could be harnessed in any practicable way, would satisfy the average requirements of 1 in 1000 of all the Earth's inhabitants.

Energy of one average lightning flash = 100 kWh = 84,000 kcal.

6.6 Electricity, the World's Most Costly Commodity

– The *cheapest* bulk product of industry, delivered at the consumer's home as a tangible, visible, storable, and harmless commodity, available day or night at any time and in reasonable quantity, is, in fact, just . . . domestic water.

The average use by the world's population is estimated at 10,000 to 20,000 kg per person per year; in the industrialized countries the price is of the order 0.01 cent per kilogram (about 0.005 pence per pound).

– The *most expensive bulk* product of industry, delivered at the consumer's home as an intangible—indeed untouchable—invisible, not easily stored and by no means harmless commodity, but again available day or night at any time and in any reasonable quantity, is, in fact, just . . . electricity.

Converting the world's annual consumption of this commodity, about 4×10^{12} kWh into its mass equivalent by means of the Einstein relation $E = mc^2$, the total weight of the electric power consumed is 150 kg (330 lb) per year. This is 0.05 mg per capita per year.

The price of this product to the consumer being 600 million dollars per kilogram (270 million dollars per pound), it is perhaps the most expensive commodity existing—500,000 times as costly as gold.

– If each individual's water consumption were delivered to him once a year by a 15-ton tank car, the weight of electricity to be delivered at the same time would be that of a tiny fraction (a few ten-thousands) of one drop of water.

1 gram = 25 million kilowatt-hours energy equivalent.

Price of the electric current at the power plant = 0.6 cents per kilowatt-hour. Delivered at home, about 3 cents per kilowatt-hour. Price of gold = 1200 dollars per kilogram.

6.7 Hidden Power

The total power under the hoods of automobiles in the Western World (including, one may suppose, cars with rear-mounted engines) is about 10 times the power capacity of all the electricity-generating stations.

6.8 Odd Bits

- If the sun were made of a mixture of oxygen and the best quality of coal, this coal would be burned up in one or two thousand years at its present rate of heat development.
- Whereas a big American car consumes 1000 kcal in the form of gasoline when driving 1 km (0.6 miles)—and a small European car half that quantity—a man consumes only 30 kcal when walking that distance.

- The total annihilation of 1 gram of coal would yield as much energy (according to the relationship $E = mc^2$) as the combustion of a whole trainload, 10 cars, each of 20 tons capacity, thus 200 tons of coal.
- Nature needed 200 million years to produce oil from microscopic sea animals, but man is using up these deposits within 200 years.

Chapter 7

Perception

I can never believe that touch, taste and tunes are more evil than the sight or the smell of flowers, which no one condemns.
(freely quoted after Charles Morgan)

7.1 How Impressions Reach Us

The speed of propagation of sound is the geometric mean of the speeds at which light travels, and with which smells diffuse in consequence of Brownian movement (of the molecules):

- Light 3×10^{10} cm per second
- Sound 3×10^{4} cm per second
- Smell 3×10^{-2} cm per second

7.2 The World of Sound [31]

7.2A A Field Full of Surprises

The science of acoustics is a field full of surprises to those unacquainted with the physics of sound, and even such familiar transmitters and receivers of sound waves as our vocal cords and our ears conceal scientific truths of which we are normally but dimly, if at all, aware.

Most sound-producing mechanisms are extremely inefficient, their output in terms of energy being only a minute fraction of the mechanical

A field full of surprises.

or other energy necessary to activate them. Consequently, most of the sounds of our daily lives—the crying of a baby, the buzzing of a bee, the hum of a passing car—all represent powers almost inconceivably small in comparison with that issuing from such sources of radiant energy as a burning candle, an electric light bulb, or an electric heater.

In view of these facts it is astonishing that the human ear is capable of detecting such low levels of acoustic energy. Under favorable conditions, the weakest audible sounds correspond to energy acting on our auditory system at the rate of only 10^{-16} watt. In fact, this "auditory threshold" energy, corresponding to a vibration amplitude as small as one-tenth the molecular diameter of the common constituents of air, is close to the energy of thermal motion under "ordinary" conditions. It is fortunate that our ears are not even slightly more sensitive, otherwise we would be aware of the constant background noise of the collisions between molecules of the air's constituents, a very trying situation! Indeed, sufficiently patient listeners might even be able to hear the grass growing.

On the other hand, the ear can endure, without damage, short-lived surges of sound 100 million million (10^{14}) times as high as the (lower) threshold value. The total output in acoustic watts of such sound sources (only a fraction is caught by our ears) is comparable to the electrical energy output of a power station serving a small city; as we shall see, the small fraction caught by our ears is something like the power consumed by a small pocket flashlight.

The eardrum and the cross section of the outer ear have a surface area of approximately 1 cm^2, so that the power impinging on the eardrums is only a very small fraction of the total power emitted by the sound source. Nevertheless, under ideal conditions of sound propagation (i.e., if no losses owing to conversion of acoustic to thermal energy took place, and in the absence of wind deflection, and reflection from the Earth's surface) the human ear could detect a sound source emitting acoustic energy at the rate of 100 watts ($\frac{1}{8}$ horsepower) at a point as far from the source as the distance between Brussels and Athens, or Boston (Massachusetts) and Denver (Colorado), i.e., about 3000 km, or 1875 miles!

7.2B A Walk through the World of Sound

Though our ideas about acceptable noise levels are generally very subjective it is true to say that a tenfold increase in the power of a sound source doubles the degree of loudness. If we found ourselves in an environment where the level of sound was at the threshold of hearing (10^{-16} watts impinging on our eardrums), and we wished to visit five different localities in each of which we should be subjected to sound levels ten times as great as the last, our itinerary might be:

	Watts per cm^2	Decibels (dB) or phons
• a meadow where the grass rustles faintly in a gentle breeze	10^{-15}	10
• the center of a deserted city park	10^{-14}	20
• a street closed to traffic	10^{-13}	30
• a large city square with its background hum	10^{-12}	40
• a vantage point some 50 yards from a busy highway	10^{-11}	50

The noise level has now increased by equal steps to 100,000 times the level at our starting point, i.e., the auditory threshold. At this level we might hear dolphins emitting sound waves of frequency between 150 hertz and 120 kiloHz as they search for their diet of fish.

Entering a busy department store, we expose our ears to a further increase in sound level, perhaps tenfold, to 60 dB, and this rises by another factor of 10 as we make our exit to a street in the throes of the peak-hour traffic, where the noise level is about 70 dB.

Just as we are inclined to think that 10 million times the threshold value for hearing is about the limit of our endurance, the roar from a train passing over a bridge just above our heads—80 dB—convinces us that we can stand at least a further exponential increase. In fact, the exhaust crackle of a powerful motorcycle a few feet away (90 dB), and the staccato pounding of a pneumatic drill across the street, each of which represents a successive heightening of the sound level by the same factor of 10, prove to leave our ears momentarily deafened but still intact at 100 dB.

By this time, we have subjected our ears to levels of sound considerably beyond what might be called the *damage-risk limit*, i.e., the noise level at which we may suffer permanent loss or degradation of hearing if exposed to it for long frequently recurring periods. Nevertheless, the acoustic power concentrated on our eardrums by the pneumatic drill (10,000 times the lower threshold value) may still be increased 10 thousand times before we experience the so-called *threshold of pain*, (10^{-2} watts per square centimeter or 140 dB), i.e., the noise level at which we immediately experience severe pain in our ears, and to which we should, for reasons of safety, never allow ourselves to be exposed for more than a few moments.

Airport ground staff, who would otherwise be subjected to this noise level in the close proximity of modern jet aircraft are instructed to wear ear protectors. Even this extreme level of sound, however, is negligibly small in terms of incident energy when compared to the power consumed by a pocket flashlight.

7.2C Full of Sound and Fury, Signifying . . . Very Little

When the extremely small amount of energy which, impinging on our eardrums, gives us the sensation of sound, is expressed in other terms the equivalent values seem almost comical.

The acoustic equivalent of the energy needed to warm up a teacupful of water to 50°C (122°F) is, by way of examples:

- 10 years, close by an aircraft's propeller (sound energy, 10^{-4} watts)

- 1000 years of travel on the underground railway (sound energy 10^{-6} watts)
- 1000 million years, listening to the whispered "sweet nothings" of one's boy—or girl—friend
- 1 million million (10^{12}) years in a quiet meadow, with no wind, listening to the birds flying by. However, this length of time is greater than the whole time for which birds, and even the Earth itself, are believed to have existed!

The sound energy developed by a large concert orchestra in the course of a 2-hour program is about equivalent to the energy used by the conductor in taking the last six steps to his dressing room. In terms of mechanical energy, even hearty applause from the entire audience is a very meager token of appreciation of the orchestra's performance; it corresponds to something like the energy expended by one member of the audience in raising an arm up to his head.

The power output of the sound produced by a man who talks for 3 hours a day throughout an average life span is roughly sufficient to heat a teacupful of water up to the temperature at which tea is normally drunk. Or, in other words if the speaker's food consumption is 3000 kcal per day, his life-long oratory will absorb the energy equivalent of a very small slice of bread, without butter!

If all the inhabitants of the world were to speak at the same moment, the total acoustic power produced would be comparable to the output of a rural power plant (0.3 megawatt).

In order to produce a level of sound approximating that of a modern jet engine operating at full thrust, the whole world would need to shout about three times as loudly as a normal speaking tone while the total volume was concentrated at a single point.

The human being is thus a poor "loudspeaker," but has a remarkable capacity for enduring extremely high levels of sound, and for perceiving extremely low levels of sound. In view of this capacity, it is apparent that the displacement of the eardrum from rest in response to the lower threshold values of sound must be almost immeasurably small. In fact, it is of the order of one-tenth of the diameter of the hydrogen molecule, i.e., one thousand-millionth (10^{-9}) of a centimeter. Meanwhile at the upper threshold, the "pain" level, sound pressure on the eardrum is about 0.5 gram (per square centimeter).

In touching on some further faculties of our ears, we compare them with our eyes as receptive organs (see Section 7.6).

7.3 Sound Pressure [32]

When an aircraft flying at a height of 15 km (about 50,000 feet) breaks the "sound barrier," a pressure wave amounting to 7 kg per square meter (about $1\frac{1}{2}$ lb per square inch) is experienced on the ground.

Assuming that the surface area of an ordinary-sized house is 150 square meters (1600 square feet), this pressure wave can be compared to the fall of a motorcar, also "ordinary-sized" and so weighing about 1000 kg, or 1 ton, on the house. However, the effect of the pressure wave is much less than that of the car, because of the slower buildup of the pressure wave compared with the impact of a falling object and its more even distribution over the roof: otherwise in these days of "sonic booms" we should have houses falling in right and left (as well as "center"!).

On the human eardrum, with its area of about 1 cm^2, the same pressure wave will cause a force of about 0.7 gram. We know that "hearing" a sonic boom is an uncomfortable, even painful, experience, and this sensation of discomfort or pain shows how sensitive the eardrum is. To be sure, this is less than the sensitiveness of the cornea in the human eye, on which a pressure of 0.2 grams per square centimeter is enough to cause a sensation of pain, but our eardrums are no less than 100 times more sensitive than the backs of our hands, and as much as 300 times more sensitive than our fingertips.

7.4 Musical Sounds in Violin Units

Musicians have as a rule an everyday language different from that of the physicists: the "decibel" is not the musicians' usual measure of the volume of tone of their instruments. For them, a more suitable unit is perhaps the sound of one of the quieter instruments, the violin. On this basis, comparative powers are about thus:

	Measure in "violins"
• Viola	1–1.5
• Violoncello	2
• Double bass	2–3
• Woodwind instruments	4–5
• Brass	10–20
• Kettledrums	40
• Cymbals	100

It must be added that an instrument played "forte" is about 4 times as loud as the same instrument when played "piano."

7.5 Overpaid Professors

In pronouncing a vowel, we use much more energy than when we pronounce a consonant; "*a*" as in "talk", for instance, needs 1900 ergs, as against "*v*" as in "vice," which needs only 1 erg.

To pronounce an average syllable, a speaker uses about 200 ergs, and a professor giving a 1-hour lecture will use about 10,000 syllables. Accordingly the total energy expended will be about 2 million ergs, which is the equivalent of 6×10^{-8} kWh. At a reasonable electricity price of 5 cents (U.S.) per kilowatt-hour, a professor's pay could be fixed at 3×10^{-7} cents for each hour of lecturing.

On the basis of energy only, professors seem to be greatly overpaid; but it is *Professor* Gamow who drew this conclusion [33]!

7.6 The World of Light

Faced with the choice of having to accept either blindness or deafness, most of us would presumably resign ourselves with greater reluctance to darkness than to silence. Social and psychological considerations apart, however, our faculty of sight is in certain respects far inferior to our powers of hearing.

7.6A The Visual Threshold

Human eyes, or rather one of them, can perceive [34] as few as 100 quanta per second of yellow light (wavelength 0.512). This, in other terms, is 3.6×10^{-17} watt incident on a pupil of area 0.5 cm^2, corresponding to a lower limit of vision of 10^{-16} watt per square centimeter, the same energy as that of sound at its lower threshold value (see Section 7.2).

The range of luminosity that allows for visual discrimination is [34] represented by 1 to up to 10^{10}, but evidence exists that the limit which one can endure is even higher than the limit for our hearing, which is up to 10^{14} times the lower threshold value.

To make the data concerning these limits of perception more readily comprehended by the imagination, consider the following experiment:

We drop a peanut (0.6 gram) from the ear/eye level to the floor (distance, 1.7 meters). If the potential energy involved—1 gram-meter—were to be wholly converted into sound or to light energy, this would have been sufficient to activate either the aural system or the optic nerve of a man to the following levels:

- during 3 million years—a time longer than the whole existence of the human race—just perceiving a sensation of sound or of light, i.e., 3 million years at the lower thresholds of both hearing and vision;
- during 1 second, experiencing sound at the pain threshold, i.e., the upper threshold for hearing;
- during $2\frac{3}{4}$ years, experiencing light at the upper threshold for visual discrimination.

Conclusion: from an energy viewpoint hearing and seeing are just (pea)nuts.

7.6B One Octave or Nine?

Taking from the field of music the term "octave" as meaning a doubling—or halving—of frequency, only about one octave of the *frequency spectrum* of electromagnetic waves (between the wavelengths 3800 Å and 7600 Å) is visible to the human eye. But, at least nine octaves of the sound "spectrum" (between wavelengths of 3 cm or $1\frac{1}{4}$ inches and 10 meters or 33 feet) are audible to our ears (see Section 2.25).

7.6C Detecting Short Intervals of Time

Our ears have another remarkable faculty, that of separating sounds arriving with a very short interval between them; this interval is in fact no more than 30 microseconds (3×10^{-5} second). This ability contributes to our enjoyment of stereophonic reproduction from a radio or a phonograph, for the twin speakers reproduce for us the tiny time intervals between the arrival of sounds from the nearer and the more distant instruments of an orchestra, as well as (of course) the differing loudness of these instruments as "heard" by the microphones. The combination of these two effects gives us the "stereophonic" impression when the sounds are reproduced.

In terms of distance, the second of two sound waves which we can just hear as separate (thus 30 microseconds apart in time) will be 1 cm behind the first. The significance of this apparently not so small difference may be illustrated by the following comparison.

If two jet aircraft race by, not far above our heads, one leading the other by 1 cm, then, to equal the ear's performance in detecting a short time interval, our eyes would have to give us clear signals as to which of the aircraft crossed some imaginary line first and how long afterwards the other aircraft crossed it.

Still another illustration may come from the rifle range. We know from experience that if two rifles are fired at the same time, though at differing distances from us, we hear the reports as two separate sounds, with a shorter interval as the two marksmen come nearer to each other. Accurate measurements have shown that we only fail to distinguish two bangs when the rifles, the sources of the sounds, are within 1 cm of each other.

7.6D Resistance Against Static Pressure

A last (minor) point in which the ear is superior to the eye is that the eardrum is only about $\frac{1}{3}$ as sensitive to static pressure as is the cornea, in the eye.

7.6E Detecting Small Differences of Frequency

There is one important respect in which the ear is inferior to the eye. The ears' capacity for distinguishing different frequencies is about 0.3% of the frequency in the middle of the audible range, thus at about 500 hertz (Hz). In more musical terms, a sensitive ear should be able to distinguish 32 different pitches within a whole tone near the middle of the piano keyboard, say from c^1 (512 Hz) to d^1 (576 Hz) with steps between each of 2 Hz.

To equal this visually, let us imagine sound waves of frequency 500 Hz to be in some way rendered visible; their wavelength will be about 66 cm. Steps of 2 Hz will correspond to a wavelength difference of about 2 mm, and, since a sharp eye can distinguish a measurement of 0.1 mm on a good draughtsman's scale, the eye has indeed a noticeably greater power of separation than the ear, considered proportionately. In the simple test just described the difference amounts to 20 times.

7.6F The Angle of Reception

Another feature in which in theoretical terms the ear is inferior to the eye is that the angles within which we apprehend impulses are for the ear, 15–20° and for the eye, 90°. However in real life, because of the different means of propagation of sound and light, this is no handicap; we can hear a car coming behind us, even though we cannot see it.

7.7 The Ear versus the Eye

Summing up the conclusions in Sections 7.2 and 7.6, we can state that:

– The ear is *superior* to the eye in two respects.
  - The audible range of frequencies is nine octaves, against only one octave for the visible range of frequencies.
  - The eardrum can withstand about 3 times the static pressure which is the limit for the cornea, in the eye.

– The eye and the ear are *about equal* in one respect:
  - The lower threshold of perception for both instruments is at the same incident energy level, i.e., 10^{-16} watts per square centimeter.

– The eye is *superior* to the ear in two respects:
  - The eye has an advantage of 20 times in separating nearly equal frequencies.
  - The angle within which the eye can receive direct signals is about 5 times the actual reception angle for the ear.

7.8 Adaptation to Light [35]

The obvious mechanism by which adaptation of the eye to light could be explained is the regulation brought about by changes in the size of the pupil. However, the pupil's diameter varies only between 2 mm and 8 mm, producing a sixteen fold change in area.

This can account for adaptation over a range of only one-tenth of the total intensity range which the eye can deal with (see Section 7.6A); the major part of this "adaptation" is attributed to changes in the sensitivity of the retina.

7.9 Seeing Is Not Perceiving

It is estimated that only 1/1000 of 1% (i.e., 1/100,000) of all the information reaching the retinas in our eyes leads to actual perception of our surroundings.

Moreover, perception itself requires appreciable time, about $\frac{1}{2}$ second. Suppose a rabbit sits on the road in front of an approaching car. If the car is 15 meters (about 50 feet) away, traveling at 100 km (60 miles) per hour, the driver's visual perception of the rabbit—half a second later—will be at just the instant that the rabbit is, unfortunately, run over.

An optical device functioning without any delay would have to be located about halfway to the moon for registration of the appearance of the rabbit to coincide in time with its regretted demise.

The reason for this delayed communication by our nervous system is that its communication lines, the axons, do not carry an electric current the way a wire does. In a copper wire the electrons travel at the speed of light, 3 million times faster than the 100 meters per second by which the impulses of the amplifying and relay system of our nerves propagate. In this respect the squid is better equipped than we are.

7.10 The World of Smell [36]

7.10A How Many Molecules Make a Man Smell?

An ordinary "sniff" amounts to about 50 cm^3 of air, and it is estimated that 2% or so of this comes into the olfactory clefts where our organs of smell are located. Experiments with the very strong odorant ethyl mercaptan suggest that about 8 molecules are sufficient to trigger an impulse in a single nerve ending, and that a sensation of smell is perceived when about 40 nerve endings are so "triggered." So, as a limiting case, only 300 molecules of ethyl mercaptan should be enough to produce a sensation of "smell," but in reality a much greater number of nerve endings than 40 are inevitably involved so that the quantity to be "sniffed" is vastly greater than 300 molecules if the very noticeable—and when used as a warning of leakage, very useful—smell is to be perceived.

At the other extreme, the highest concentration of ethyl mercaptan which can be tolerated (pain threshold) is 10^7 times the foregoing, thus 3×10^9 molecules for each nerve ending, a number about the same as the present population of the Earth.

As to differences, the "just noticeable difference" (JND), the difference in odor-intensity which can just be perceived for ethyl mercaptan is such that there are 30 JND's between the lower (perception) threshold and the upper (pain) threshold. This means that for each JND the number of molecules necessary for one olfactory nerve ending increases by 100 million, a number equal to about half the population of the United States.

7.10B "His Master's Odor"

Human scent can readily detect among other substances, various "fatty acids." Our noses are so sensitive that if we sniff air in which the concentration of butyric acid (molecules) corresponds to the ratio of one man to the whole world population—1 in about 3×10^9—we can detect the odor. There are indeed materials (for example α-ionone, with the odor of violets) to which our noses are 25 times as sensitive as they are to fatty acids.

But these achievements are modest compared with what a bloodhound can do. If only 1/1000 of the 16 cm^3 of moisture which the underside of a man's foot produces daily diffuses uniformly through the sole of his shoe, his footprint will still contain 2.5×10^{11} molecules of fatty acid. This is enough to impart "threshold concentration" for the bloodhound to 30 cubic meters (40 cubic yards) of air, and it has been calculated that under normal conditions of temperature, wind, and rain the footprint can remain noticeable—to the bloodhound's nose—for 24 hours.

To take over the bloodhound's task, a man would need to increase his nose's sensitivity to these fatty acids 1 million times. On the other hand, some sort of fish beat the bloodhound in smelling sensitivity by a factor of 100. Operating on a "somewhat" different scale concerning distances, these fish would be able to smell the presence of one particular man on a cluster of 20 million earth globes, all with the same population density as ours.

Threshold of smelling fatty acids by a bloodhound: 1 in 10^{15} grams.

7.10C Like Ships That Pass in the Night

From the foregoing one would perhaps conclude that dogs are very "efficient" trackers; however, as we shall see, this is only true within certain limitations.

Perspiration comes of course not only from the soles of the feet, but from the whole body, at a scale estimated at a total of 1 liter per day. In this is contained 2.5×10^{16} molecules of aliphatic acids (including butyric acid), a quantity sufficient to impart scent-threshold concentration for a bloodhound to the air content of 150 "small town" gasholders, holding say 20,000 cubic meters each.

Supposing that evaporation of these acids took place instantaneously, any one of us men on an afternoon stroll would be surrounded by an immense cloud of "scent," constantly renewed, about 2–4 km across. Our dogs should therefore be able to smell their masters at quite long range—up to 1 or 2 km (but presumably all the other "masters" in the same neighborhood too).

But this possibility is limited by the slow and incomplete evaporation of the "scent" acids; the boiling point of butyric acid is 162°C (325°F), and in fact we masters get rid of most of our scent under the morning shower bath.

Tracing by scent in footprints is much more efficient because of the slow evaporation of the "scent" acids—perhaps we can liken our bodies to ships that pass in the night enveloped in a cloud, leaving a clearer trace in our "wake," that is our footprints, in which the scent may persist for hours after we have passed by.

7.10D Two Kinds of "Aroma"

A young chemical student, who should have known better than so to misuse a whisky bottle, filled it up with water, a little brown coloring matter, and, on about a half-and-half basis, ethyl mercaptan, one of the strongest odorants known, a stuff which he had just made in his laboratory. The curious feature of this substance is that it differs from (ethyl) alcohol—the active principle, so to say, of whisky and much else—in only one of the nine atoms which constitute its molecule: $C_2H_5\mathit{S}H$ instead of $C_2H_5\mathit{O}H$.

The bottle duly found its way to the shelf behind a bar, and when the next customer's turn came, the aroma of his "wee drappie" was so unusual that he dropped his glass. Next day part of the town could smell that something odd had happened. Even half a mile from the scene of disaster, a normal "sniff" (50 cm^3) contained about 10^{10} molecules of the mercaptan, which is the threshold concentration for perception by man.

7.10E Detecting Food

The threshold below which animals do not detect food varies very greatly. For example, most bees will accept a 20% sugar solution, but usually refuse a concentration of 10%.*

The threshold of perception is determined, in the case of bees, by first starving the creatures, and then giving them the choice between solutions of sugar in water of different concentrations.

Although the bee nourishes itself almost exclusively on sugars, it is ill equipped to find them, its threshold of perception being rather elevated (3% concentration of sucrose). It is not clear why in man, for whom sugar is only one of several important foodstuffs, the threshold of perception of sucrose is seven times as low.

The minnow, which tastes with its whole skin, has a threshold of perception 400 times more sensitive than that of the bee, while the Red Admiral butterfly, with taste organs in its feet, excels the bee's capacity by no less than 800 times.

7.11 From the Giant's Point of View [36]

An anatomist who makes a microscopic study of the sensory equipment of the nose is like a giant a mile high who wanted to study the communication network of London or New York.

He could begin by imbedding the whole thing in wax, and then cut thin slices out of it in various directions. If his technique was good enough, he might get the thickness of the slices down to about 3 meters (10 feet) and with the best optical devices he might contrive to view the slices from the equivalent of a hundred meters distance. He might even be able to do some dissection of the slices, but his instruments would be no more delicate than a baseball bat or a broadaxe. By using colors that would soak into plaster or concrete but not into metal or glass, he could trace the general outline of the various structures from one slice to the next.

In another kind of experiment, our giant could explore various civic functions and activities by destroying a traffic artery here or some

* In the case of the bee the reason for its refusal may be that solutions of low concentration would not yield a honey that would keep in the hive through winter: the sugar is the food, but at the same time the antifreeze. Diluted solutions would also mean uneconomical transportation.

other center of activity there, and watching for signs of congestion or decay in nearby areas.

From all this he might eventually come to have a fairly good, rather broad understanding of the communications system of a modern city. He would probably never know the detailed structure of a dial telephone, but he would be pretty sure there was such a thing. He would realize that besides the telephone system there are postal and parcel delivery systems, and he would at least partly disentangle the specialized alarm systems that send warnings of fires and violent crimes to the appropriate action centers.

Chapter 8

The Biological World

The biologist is full of hope but his friends are full of fear . . . seeing his last achievements.
(freely quoted after L. Sedov)

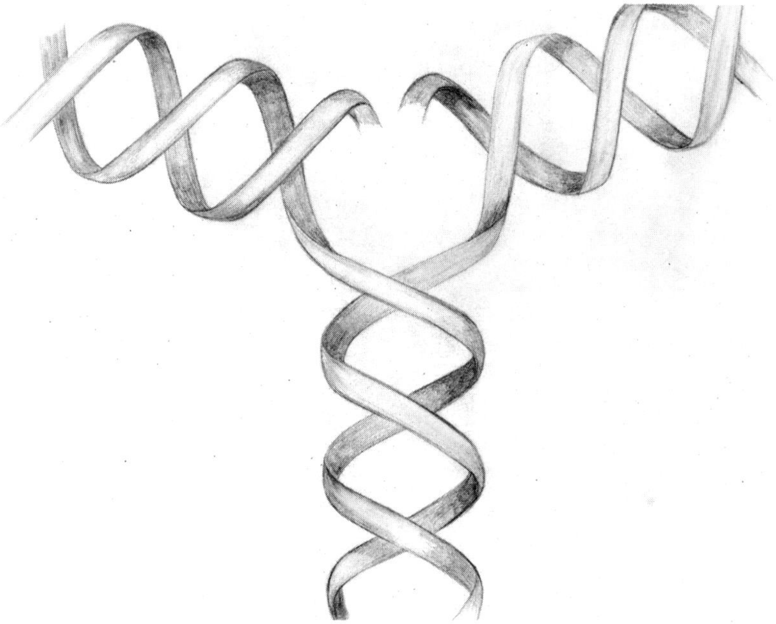

The beginning of life?

8.1 To Begin With

The number of female egg cells, amounting to 1 liter (1.5 pints), necessary for the next generation of mankind could be contained within one of all these women's hats. The corresponding quantity of spermatozoa required for fertilizing these eggs would occupy a volume no greater than a peppercorn.

The weight of a human female's egg cell at the time of conception being about 5×10^{-5} grams, its growth into a human adult means an increase in weight of 30 thousand million times, 10 times the numerical increase from our common ancestor Eve to today's world population.

8.2 Building Life [37]

The organism which causes the disease "pleuropneumonia," of a size intermediate between bacteria and viruses, is considered to be the simplest cell capable of independent self-duplication, one of the main characteristics of life. The information needed by such a cell to carry on its existence and self-duplication is thought to be—in computer language—5 million "bits," about one-seventh of the information contained in the *Encyclopaedia Britannica* reckoning 1 bit from each word.

In laboratory synthesis, the most complex molecule that has been produced so far is thought to have information "in" its structure amounting to 2000 bits, corresponding—again at 1 bit per word—to about 5 pages of a paperback.

So, as far as "built-in" information goes, more than a thousandfold improvement in technique seems to be needed before the chemist can make an attempt at synthesizing a living cell.

8.3 Astronomical Progenies

Professor Multiplier, at the beginning of his lecture, performs a simple experiment with a sheet of paper. This paper is 0.1 mm (0.004 inch) thick, and the Professor folds it, first in half, then in quarters, and so on, showing that each fold can be made in about 1 second, say a rate of 50 folds per minute. He asks his class how thick the folded paper would become, if indeed he had continued the folding for a full minute.

Although the students were well trained and experienced in mathematics, they could hardly believe that 0.1 mm $\times 2^{50}$, the thickness of the folded paper after 50 folds, is somewhere about $\frac{2}{3}$ of the distance from the Earth to the sun, that is, more than 110 million km (70 million miles).

The Professor remarked, "remember that this is only 1 minute's work in the manner described. I would also call to your attention that with only one more second's effort, you would find yourselves well beyond the sun."

"This simple example shows the fantastic increase as numbers build up in geometric series, that is exponentially, just as happens in the multiplication of living matter by the process of cell division."

"We first take—as a modest example—the 4-inch long giant African snail, which produces about 150 eggs four times in the year. A specimen 5 years old could have produced progeny numbering something over 16×10^{15}. To give an idea of the real meaning of this, I am forced," said the Professor, "to use an astronomic scale of distance; for placed head-to-tail, this progeny would reach to the sun and back again 5000 times, a distance of 2 light-months."

"If I request one of you to count this vast number of snails, counting them at the rate of 1 per second would take you as long as the time which has elapsed since the mid-Cambrian period, 500 million years ago, when most of the fossils now found were being deposited."

"But, as I said," continued Multiplier, "that was a very modest example. The cholera bacterium, with a volume of $1\mu^3$ (1 thousand-millionth of a cubic millimeter, 10^{-9} mm^3) divides every 30 minutes to produce two complete new entities. At this rate of reproduction, only 28 hours would be needed to build up a progeny equal in numbers to the snails' 5-year effort; indeed, disregarding mortality for the moment, as well as other possible disturbing factors, the progeny of a cholera bacterium would, after 6 days, occupy a volume equal to that of the Earth."

"Happily," concluded Professor Multiplier, "such rates of increase are found only in small creatures. If our own species, man, were to multiply itself at the rate achieved by the cholera bacterium, within 50 hours our descendants would be standing shoulder to shoulder over all the land area of the Earth, that is at a density of 4 per square meter (about 3 per square yard). In the next 30 hours, thus after 80 hours in all, our descendants would fill the whole *volume* of the Earth; where they could find even 'standing room only' remains an unanswered question."

8.4 The Library of Heredity

8.4A Building Up Man's Library

The basic carriers and transmitters of heredity characteristics are giant molecules of deoxyribonucleic acid (DNA), which form pairs of interlocking spirals along which variations in the chemical structure are a "code" whereby the heredity characteristics are passed on. At cell division, the spirals separate, each forming a "model" for its new partner.

The human body is comprised of some 2.5×10^{13} cells, each containing 25,000 DNA molecules having a molecular weight of at least 120 million with their spirals of about 5000 turns.

The information stored within this system will be compared to a library with books, containing as we will see, extremely long (DNA) sentences.

• *A letter.* The building blocks of the DNA molecules, the *nucleotides*, bear the letters, A, C, G, and T: so we dispose of a four-letter alphabet. To give an idea of the composition of such nucleotides, two of them are mentioned:

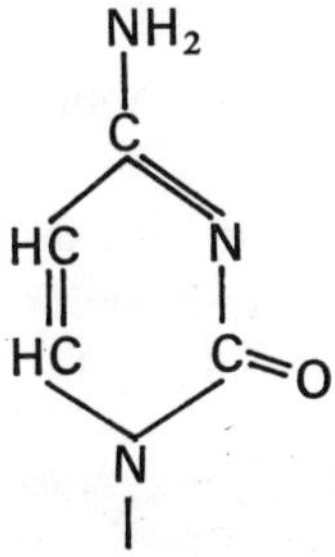

letter C
(cytosine)

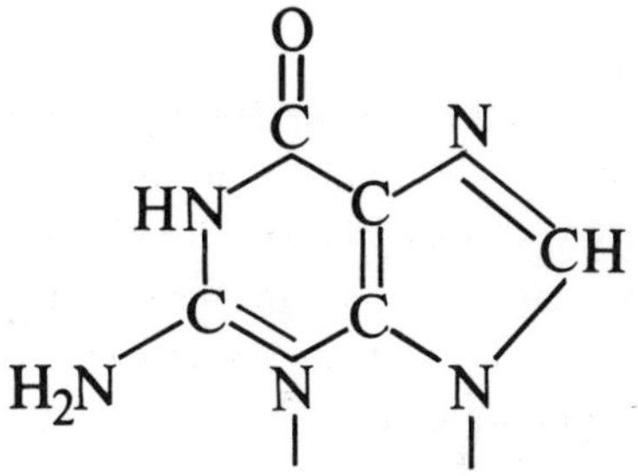

letter G
(guanidine)

A C G T

4 nucleotides

• *A word.* On the DNA molecules the heredity coding for the synthesis of proteins (see Section 8.4B) is effected by the presence and location of any three of the building blocks (letters), for instance A C T. We will call this a code word in the library.

a code word in our library

its biological equivalent

● *Part of a sentence.* Parts of the long DNA sentences (DNA molecules), the *genes*, consisting of 100–300 code words determine one characteristic heredity property (or group of properties) of man, animal, or plant. This word count is, for a man, comparable to $\frac{2}{3}$ of a page of a book, such as *Data: Mirrors of Science.*

part of a sentence

200 code words
($\frac{2}{3}$ page)

● *A full sentence—a DNA molecule*—carries further instructions about the synthesis of other proteins. In man, the most complicated being, a DNA sentence carries about 1000 genes; it is comprised of 370,000 nucleotides, which means about 120,000 words, as many as in 2 copies of our book *Data: Mirrors of Science.* In less complicated beings this number is smaller.

a full sentence

1.2×10^5 code words
in 2 copies of this book

● *A bookshelf.* The DNA molecules are grouped into 46 *chromosomes.* In our library we can liken them to bookshelves, each containing some 500 DNA's, or 1,000 copies of our book. Or, if you prefer, 7 times that of the English Bible.

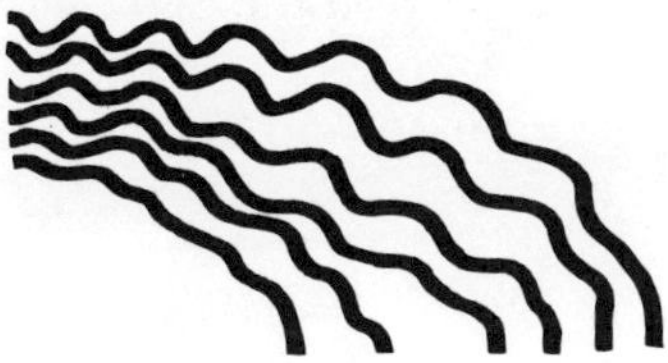

The bookshelf

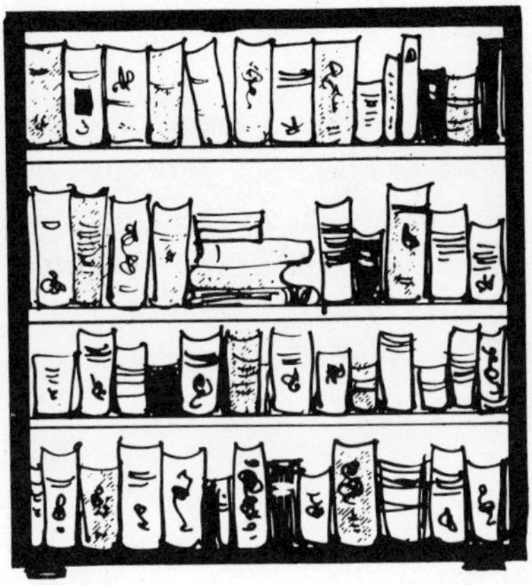

1,000 books, containing 6×10^7 code words.

● *A reading room* in the library, a *cell* in the human body—for man equipped with 46 book shelves—is a complete entity of a living creature. Altogether the room contains 46,000 books, i.e., roughly 3×10^9 words: one word per living man today.

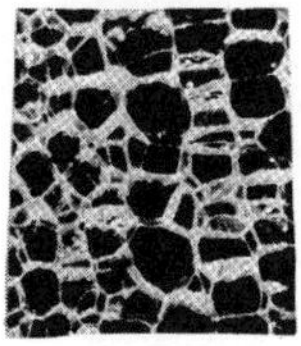

Cells

A cell: 46,000 books containing 3×10^9 code words.

● *The library—man* himself—built up by 2 to 3×10^{13} cells is a monotonous repetition of say 2.5×10^{18} of these reading rooms, all exactly the same, with the same books, containing the same sentences. But, since brain cells differ very much from, for example, liver cells, it can be concluded that in each cell only certain parts of the sentences are effective in determining the course of protein syntheses or other functions.

All the reading rooms together, each 13 meters long (height of the shelves: 12 books = 3 meters), placed in a long, long row, would form a corridor stretching 10 million times around the Earth or again—if you prefer—it would cover 2000 times the distance between the Earth and the sun.

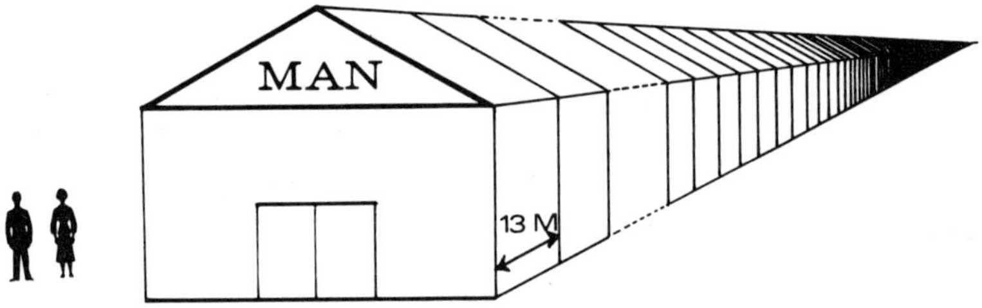

A library with 2.5×10^{13} identical reading rooms, 10 million times around the Earth: 6.8×10^{21} code words.

8.4B The Information Capacity

As mentioned before, the code words determine which amino acids, the elements that constitute protein, shall be synthesized in the body's cells. These proteins are not only the basic building material in enzymes which catalyze the body's vital processes but also in structural elements such as cell walls, muscles, and nerves.

Choice and arrangement of three out of four letters as indicated gives the possibility of $4^3 = 64$ different code words, but in Nature we find that the permutations and combinations result in only 20 different amino acids. Nevertheless the number of possible variations is overwhelming. A DNA chain, only 100 units long, could be arranged in 10^{130} different ways [73]. This number is 10^{45} times larger than the number of atoms in our solar system! And the DNA chain in a man is not 100 but, as we saw, 370,000 units (nucleotides) long. The number of possible variations in arrangement becomes so great that no realistic comparison can be found for it (see also Section 8.9). How can we ever hope to calculate the probability that another Chopin or another Churchill will be born, identical with his predecessor?

Considering the enormous number of the units in which human heredity is stored, it must strike us as marvelous that relatively so few spontaneous mutations, that is mistakes in the different syntheses which precede cell division, occur. The proportion amounts to about 1 in 1 million divisions of the cells which constitute our bodies (see, for such "spontaneous mutations," Section 8.12).

In simpler creatures the information capacity of the DNA sentences is considerably smaller. Whereas in an average cell of a mammal the uncoiled DNA molecules, if placed end to end, would form a thread, 1 or 2 meters long, the DNA in a large virus if uncoiled and placed end to end would form a thread about 0.1 mm long. The chains in a mammalian cell are so thin that if one were to be laid crisscross on the head of a pin, only 0.5 per cent of the available area would be covered.

8.5 The Genetic Staircase

As pointed out in Section 8.4A, the coiled-up human DNA molecules of molecular weight 120 million or more, have spirals of about 5000 turns. When paired, the spirals are separated by a distance of 20 Å, by the agency of chemical side chains which may be compared to the treads of a spiral staircase, each 3.4 Å high and 10.8 Å broad.

In the chemistry of living matter, the carbon atom is by far the most frequently occurring "building brick." It has a diameter of 1.5 Å; in

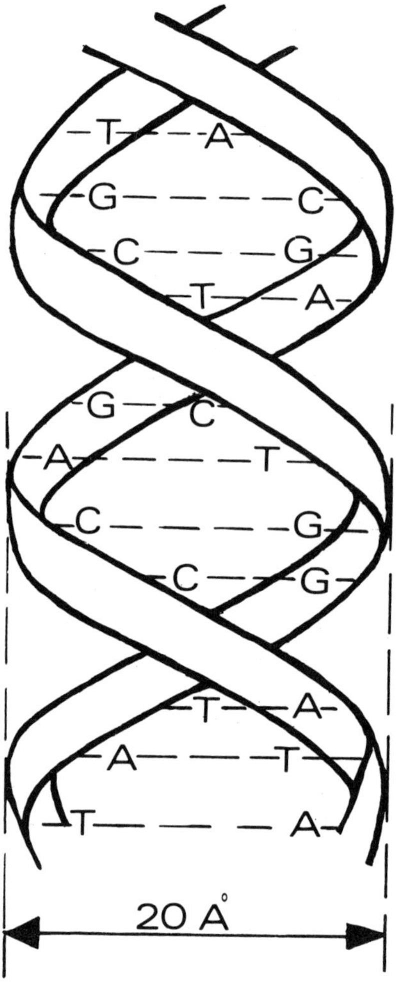

Paired spirals in DNA, to be compared to the treads of a spiral staircase.

comparison with our ordinary brick for houses and the like—sized about 20 cm × 10 cm × 5 cm or a little larger—it is equivalent in volume to a cube of side 10 cm.

Now, if we magnify the carbon "bricks" in DNA to the size of ordinary bricks we should find a structure so high as to reach above the troposphere, that part of the atmosphere where "weather" phenomena have their place; in fact to a height of about 10 km (about 33,000 feet), so well above Mount Everest.

With a breadth of 70 cm (about 28 inches), and a rise at each tread of 23 cm (about 9 inches), we have just such a spiral staircase as we find in the turret of a medieval castle. With a few rests for refreshment and even sleep, any of us might—in the end—reach the top of a "brick" DNA molecule: person after person on this narrow staircase.

8.6 The Robe of Heredity

A full-grown man or woman has at least 10^{13} cells in the body, and in each of these cells the total length of the uncoiled deoxyribonucleic acid spirals is about 1 meter.

Imagining these spirals as a single long thread, we could form from it a dignified robe for an average member of the human race. Though it would be 10,000 threads "thick" at any point, this would amount to only 0.02 mm, the thickness of a very light-weight nylon stocking.

Another way of emphasizing the enormous total length of DNA molecules in our body cells is that, stretched out straight, they would form a thread reaching back and forth from the Earth to the sun 33 times.

Distance from Earth to sun $= 1.5 \times 10^8$ km (9.3×10^7 miles).

8.7 A Remarkable Setting Machine [38]

An *Escherichia coli* bacterium divides under optimum conditions in 20 minutes or roughly 1000 seconds. In that time it synthesizes its own (dry) weight of 1.5×10^{-13} gram of protein.

If we take the molecular weight of the large proteins involved to be 100,000, then the weight of one molecule is 1.6×10^{-19} gram and the number of protein molecules synthesized per second per cell is 1000 in round numbers.

Since each molecule involves the correct ordering of something like 1000 amino acids, it is clear that a very remarkable synthetic method must exist in order to arrange in each cell every second 1 million amino acids according to *the* predetermined code existing in the cell DNA (see Section 8.5). Which man-made setting machine can do this job?

It has been estimated that even in 100 million replications of the genes of a typical bacterium there is only a 50% probability of one being altered (resulting in a mutation).

This leads to the thought that with a replication mechanism of this high efficiency, mutations might become so infrequent that a species could become trapped in an evolutionary "dead end."

8.8 Whirling Stems [39]

In the process of biological replication in the growing cell we now believe that the double helixes constituting molecules of deoxyribonucleic acid (DNA) have to unwind, and in this the central cores or "stems" have to turn also. It appears that the unwinding which precedes replication takes 20 minutes, and if this is indeed so, it means that the rotational speed of the central "stems" is about the same as that of the rotor of an ultramodern centrifuge, that is 50,000 revolutions per minute (see Section 10.15)—that is, 1000 revolutions for each "tick" of a grandfather clock.

8.9 Anemia, and a Printer's Error [62]

A molecule of hemoglobin, the constituent of blood which is responsible for its striking red color, is made up of more than 500 amino acids of 20 different types. The occurrence and location of these amino acids can be in any possible combination, and the number of arrangements, and thus the possible "mistakes" in the process of synthesis which the arrangement of these acids determines, is in fact greater than 10^{600}, an unimaginably large figure when we remember that the perceptible universe is considered to contain [39] about 10^{75} atoms—some conservative authorities favor an estimate of 10^{70}!

Sickle-cell anemia—a kind of anemia characterized by the form of the (red)cells of the blood when in a low partial pressure of oxygen—had been called a "molecular" disease, being an inherited alteration of only one amino acid in each half of a hemoglobin molecule, or more precisely one alteration in the sequence of bases in the chromosomal DNA mould of a sufferer from this disease. It is one of the serious errors—comparable to one wrong letter in printed material, a "printer's error"—which Nature occasionally makes; in this case the usual result is early and untimely death of the victim.

Based only on probability considerations, the chance of sickle-cell anemia ever having occurred, even if we suppose 1 in 10^{600}, is a tiny fraction of the chance that a space traveler would, during the same period, have collided with one particular atom somewhere in the universe.

8.10 Lethal Doses

The lethal radiation dose for Mr. Victim, who weighs 70 kg (155 lb), is 700 rads, which means an energy input of 5000 million ergs (120 calories). Consequently, radiation appeals to Mr. Victim's enemies as a very efficient weapon, the energy required to effect his demise being merely equivalent to:

- raising his body temperature by 0.002°C, or
- causing him to fall from a table, 1 meter (3 feet) high.

Notwithstanding this efficiency in killing from an energy point of view, it requires a special technique because every living person would have to "fire" 1 million quanta of the most concentrated kind (γ-radiation of 1 MeV) at Mr. Victim before he would be certain to succumb.

or, expressing it in other terms:

- if the number of molecules per gram in an animal is 3.3×10^{22} (as in its major constituent, water) then death occurs when one molecule in 35 million is affected by radiation. It is as if the whole population of Spain would die if one member were killed by a bull.
- for a 70-kg man, if the molecular weight of the affected molecules was 100,000, then the weight of damaged material causing death would be 1 gram.

8.11 Radiation Dangers [64]

On the average, an inhabitant of the United States living at sea level absorbs about 0.15 roentgens* per year into his skeletal system from natural sources. Of this, only 20% comes from worldwide fallout consequent on nuclear experiments, etc., which means shortening our lives by 2 days. To assess the scale of this possible danger to health, so much discussed, let us compare it with other known dangers.

* One roentgen unit is the quantity of absorbed radiation, as defined by the ionization produced in dry air. One roentgen will produce ions carrying 1 electrostatic unit of electricity in 1cm³ of dry air.

- It is estimated that 10% overweight reduces one's life expectancy by 1.5 years.
- A steady smoker—1 packet of cigarettes per day—is estimated to shorten his (or her) life by 7 years.

One can see from this, that the danger from fallout is the same as that of smoking 1 cigarette every 2 months.

Cosmic rays, gamma rays from the ground, radium in drinking water, and living in a concrete house (even concrete is a little radioactive) are all more injurious to men than present experimental nuclear fallout, each corresponding to smoking about 6 cigarettes a month.

Due to the Earth's natural level of radiation each individual receives only about 20,000 quanta per second, at which rate it would take over 5000 years for a lethal dose to be built up.

8.12 Mutations by "Sparks from Heaven" [40]

What we call "cosmic radiation" (see Section 8.11) consists mainly of protons moving with immense velocities, 1000–2000 km per second, through outer space; that is, at some 5000 times the speeds with which modern jet aircraft pass over our heads, albeit so far up as to be almost out of sight! When these high-speed protons enter the atmosphere, atoms of oxygen or of nitrogen—or any other gas—which they may strike are shattered, producing lighter particles like mesons and electrons, and X-rays. These particles and radiations collide in their turn with other atoms which then also disintegrate and so a vast "shower" of secondary particles results. In fact, one such "spark from heaven" of energy, say 10^{18} electron volts, can produce a shower of secondary (and tertiary, etc., to be exact) particles and radiation covering many square miles of the Earth's surface and numbering more than 300 million.

Although some cosmic particles have these immensely high energies (see Section 4.9), the total energy of this radiation is small, as discussed in the previous section; even outside our atmosphere it is only $\frac{1}{3}$ of the safe long-time radiation "dose." The stream of energy which we receive from cosmic sources is of the same order of magnitude as the light coming from the stars (excluding, of course, the sun).

But locally, this radiation energy can be intense enough to affect man, causing so-called spontaneous mutations. The particles themselves are much too energetic; their impact would destroy a whole gene, and it is in fact sufficient for a high-energy particle (or radiation)—energized as we have seen by cosmic radiation—to come only into the neighborhood of a gene to bring about the sort of minor displacement in a DNA molecule which results in a "mutation" as we know it.

It is easy to calculate that the 300 million secondary (etc.) particles produced by the activity of only one "10^{18}-electron volt" cosmic particle means a distribution of at least 800 particles per square meter every 24 hours on the Earth's surface, so that a standing man is being hit every day by 100 "high-energy" particles (and even more if lying prone); so there are 100 chances each day for a mutation to be induced.

What happens now in densely populated cities? An area comparable to New York City (1000 km^2) is hit once every day:

- by 3 showers, each due to one 10^{20} eV cosmic particle
- by 3000 showers, each due to one 10^{18} eV cosmic particle
- by 70,000 showers, each due to one 10^{17} eV cosmic particle
- by 5 million showers, each due to one 10^{16} eV cosmic particle

All these showers together produce for each of the 10 million inhabitants of such a city every day and night at least as many chances for a mutation as if each of his fellow citizens had fired one particle at him during that day.

Need we be surprised that "spontaneous" mutations have indeed played a role in the rise of man from microbes or algae?

8.13 Radioactive Man

Neutrons, present in Nature as a result of cosmic radiation on the material of the Earth and its atmosphere, may in turn strike nuclei of the element nitrogen, forming the radioactive isotope of carbon, ^{14}C. Radioactive carbon, in its turn, has a definite rate of distintegration, so there results a constant, though very low, proportion of ^{14}C in the Earth's atmosphere.

Combined with oxygen, ^{14}C forms radioactive carbon dioxide, which enters plant structures through the agency of chlorophyll and finally reaches animals at the end of the "food chain." When an animal dies, however, it takes up no more ^{14}C, and so by natural disintegration the amount of this isotope of carbon in the body gradually falls. By determining how much is still present at any time, an investigator can estimate how much time has elapsed since the death of the animal, up to many thousands of years.

However, during the lifetime of the animal, there is a fairly steady proportional relationship between ^{14}C and the normal (stable) isotope of carbon, ^{12}C, that is 1 in 1000 million, which leads to the conclusion that per gram of normal carbon, ^{12}C, about 15 atoms of ^{14}C disintegrate each minute.

The average human body contains about 15 kg of carbon, so every minute there are about 2×10^5 disintegrations of the ^{14}C isotope. This produces radioactivity at the rate of about 4×10^{-3} rad per year. But a man can receive a total "dose" of 700 rad during his lifetime, so that 150,000 years could elapse before a man is in direct danger from his own radiation.

However, damage to the DNA molecules constituting the vitally important genes by "own" radiation is not entirely improbable, seeing that the energy released at each disintegration is 155 keV and that in the body the total number of disintegrations is around 200,000 per minute.

8.14 Protection against Radiation

We can protect ourselves against radiation of different kinds simply by shielding ourselves with paper of appropriate thickness:

- Against α-particles (helium nuclei) of 5 MeV energy a page of this book, or some centimeters of air would suffice.
- Against β-particles (electrons) of high energy as produced by chlorine-38 the entire book thickness would be a scarce shielding.
- Against fast neutrons, as 10 meters (30 feet) of paper, some 150–200 volumes of *Encyclopaedia Britannica* would be as good a shielding as 6 meters (20 feet) of water, the shielding now in use.
- Against γ-rays (electromagnetic waves) of 1 MeV energy from a high-activity source (e.g., 500 curies), 300 volumes of the encyclopedia mentioned could be tried. But neither this stack of paper, nor 1 meter (3 feet) of a lead shielding with a con-

siderable safety margin, would assure that any person, who worked at the source for 40 hours a week, for 20 years, would spend the rest of his life without any signs of radiation disease.

8.15 Living under Extreme Conditions [41]

The larvae of a small fly, a bloodworm, can withstand temperatures as high as boiling water or as low as liquid air (−190°C). But among the hardiest of microbiological ventures are the spores of microorganisms which are able to survive the intense cold of liquid helium (−269°C). Some spores are very resistant to chemical disinfectants and are unaffected, or even survive better, in a high vacuum.

Thiobacillus derives its energy from the oxidation of sulfur to sulfuric acid and can grow in solutions containing 3% of it. At the other extreme are bacteria adapted to highly alkaline conditions (pH = 11).

Higher organisms are rather sensitive to radiation. A dose of a few hundred roentgens will kill most mammals or plants and a dentist will restrict his local X-ray dose to a few roentgens. Yet bacteria and blue-green algae grow in the radioactive cooling water of nuclear reactors. They can resist about a million roentgens: that is a dose that would kill more than 1000 men (lethal dose for a man: some 700 roentgens); see Section 8.10.

8.16 Man—One-quarter Horsepower—Competing with Machines [42]

The biological routes whereby energy is generated from food differ, fundamentally, from the high-temperature combustion—associated with the use of some form of heat engine—ordinarily applied by engineers.

Living matter utilizes a chemodynamic system in which the chemical energy of food yields work, without the intermediate steps of "significant heat production" and "use of this heat energy," with efficiencies up to 100%; taken as a whole, a man's efficiency for physical work, based on his food intake, is about 25%.

Supposing, however, that the human body worked as a "classical" heat engine, the efficiency is then determined by Carnot's formula:

$$\text{Efficiency} = \frac{T_1 - T_2}{T_1}$$

Here T_2 means, for the heat engine, the temperature of the cooling water; for a man, this minimum temperature will be 37°C (310° absolute, or Kelvin). From this, it follows that to equal the chemodynamic efficiency of 25%, the highest temperature in the body, T_1, would have to be 415°K, that is 142°C. At this temperature our body fluids would be well above the boil and our albumen would instantly coagulate.

In fact, the chemodynamic process in the human body operates by quite a different kind of system, the actions and reactions of enzymes. The characteristics of these lead us to the idea that the human body is basically an "isothermal" engine of about $\frac{1}{4}$ horsepower (hp).

● That we are "isothermal" engines is shown by the fact that human temperature varies by only about 1.2°C (2.5°F) around its average value of 37°C, 98.5°F. The wearing of thick or thin clothing has little influence on this; regulation of the temperature is in fact effected by breathing, and by more than 2 million sweat glands—producing, in a sense, the body's "cooling water."

● That we are basically engines, or motors, of about $\frac{1}{4}$ hp (that is something like the power used in a small electric vacuum cleaner) emerges from the consideration that a man's energy output is in round figures:

- at rest 100 kcal per hour = 115 watts, or 0.16 hp

This corresponds to the power consumed by a medium-size electric lamp. We must admit, however, that this $\frac{1}{4}$ hp man can achieve remarkable things, namely:

- in heavy work 1000 kcal per hour = 1150 watts, or 1.6 hp, equivalent to the power consumed in a medium-size electric heater.

This heavy output of 1000 kcal per hour corresponds to the real-life activity of working steadily at the job of plastering a wall. There are much harder jobs, in the physical sense, such as climbing a ladder with a heavy load, but these a man can do only for a short time.

However, as measured in the laboratory, the actions of "plastering" should require only 200 watts (about $\frac{1}{4}$ hp), and this is our basis for terming man the like of a $\frac{1}{4}$ hp engine; the remainder of the 1000 watts needed by the plasterer is lost in unnecessary movements and actions in his body.

So the plasterer's efficiency is only 20%, at first sight a low figure, but the engineer, under the aegis of Carnot, has to build a formidable machine and operate at quite high temperatures, as in internal-combustion engines, to achieve a higher efficiency. In a good modern steam engine not more than 10 to 20% efficiency can be expected; diesel engines do better at up to 40%; but man, the isothermal engine, can quite often record 25% under favorable circumstances.

Now, what is the fuel supply which makes all this possible?

Each year, vegetable growth extracts from the atmosphere some 200,000 million tons of carbon, which is $\frac{1}{3}$ of the atmosphere's total carbon content. However, human needs for warmth and energy are covered by only 2% of the amount of carbon used by the vegetable world; for each man, modest in the face of the vegetable world, the weight of a car—1500 kg—in carbon from the atmosphere will satisfy his energy needs.

8.17 Chemodynamic Processes [42]

The physical phenomena underlying the energy conversions mentioned in Section 8.16 are reactions of adenosine triphosphate (ATP)

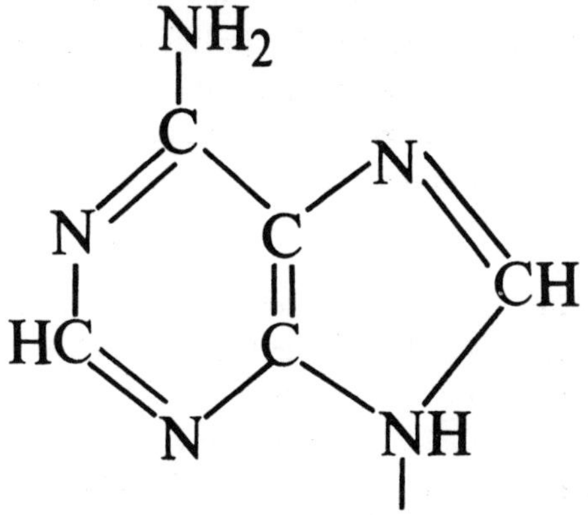

Formula of adenine

in the cells of the human body, of which the number 10^{13}, as an average for a voluminous person, is no less than 3000 times the number of the Earth's inhabitants.

On the one hand, ATP is formed by enzyme action from food; on the other, it is decomposed by hydrolysis. It is estimated that the amount of ATP so formed and decomposed in the course of 1 day in the body of a man weighing 70 kg is no less than 85 kg, thus more than the body weight itself. Indeed, it might be stated that a man's gross production of ATP during his whole lifetime would be enough to load 100 20-ton railway cars—2000 tons in all.

Yet the amount of ATP actually present in one human body at any given time is only 35 grams, for each molecule is transformed 2400 times each day, and its average life is less than 1 minute.

8.18 Gullivers and Lilliputs [43]

8.18A Motor Performances

The "cube/square" law expresses the relation whereby a process associated with the volume of a body is dependent on the cube of its linear dimension (d^3), whereas one associated with its area is dependent on d^2.

Thus, with equal rates of metabolism per gram of body weight, a larger animal has more difficulty in disposing of heat than a smaller.

The "cube/square" law not only is valid for exterior surfaces but applies also to interior surfaces of bodily organs, for example the muscles. Consequently, as a rule, the physical performances of larger animals calculated per unit of body weight are below those of smaller animals, because their muscles are relatively thinner: the results of this are plain to see in the animal kingdom.

"*Beasts of burden*"

- A horse can carry about $\frac{2}{3}$ of its own body weight
- A full grown man can carry about once his own body weight
- A little boy can carry about up to twice his own body weight
- A grasshopper can carry about 16 times its own body weight
- A bee can carry about 24 times its own body weight

The world record for weight lifting by man is 2700 kg (6000 lb) thus about 40 times the lifter's own weight, but the great rhinoceros beetle achieves a record of 850 times its weight. If an elephant's powers were on this scale he could lift a light battle cruiser!

Draught Animals

When it comes to hauling a load—on a wheeled carriage—the same picture emerges:

Certain beetles, weighing only 2 grams (30 grains) can pull a heavily loaded toy car weighing 175 grams (6 ounces); the corresponding achievement for a man would be to move a loaded truck weighing 7000 kg (16,000 lb), that is 100 times his own weight.

However even this record is soundly beaten by the edible snails to be found in a French kitchen, or by the bee, for these creatures can move loads equal to 200 and 300 times, respectively, their own body weight. A man's achievement corresponding with that of the bee would be to move off with a fully loaded railway car, 20 tons in all.

Jumpers

The high jump and the long jump are achievements which in a similar way depend on body dimensions as discussed above, modified, of course, by the anatomical features of the legs.

The showpiece in this field is of course the flea, which can do 32 cm (1 foot) as long jump and 20 cm (8 inches) as high jump. The corresponding performance by a man would be:

- long jump 210 meters (230 yards)
- high jump 135 meters (460 feet)

which would clear the length of one side and the height of the Great Pyramid.

8.18B Metabolism

In Jonathan Swift's well-known book we read that the royal mathematicians calculated what the Lilliputians—creatures 15 cm (about 4 inches) in height, or about the size of a large mouse—would have to do to provide a good night's rest for the man-sized Gulliver, whose height was about 1.8 meters (6 feet).

Their conclusion was that Gulliver, who was, for the Lilliputians, what a whale is to us, would need a bed with 144 times the area of one of theirs, equipped with 12 layers of Lilliputian mattresses, thus 1728 in all, for a comfortable night's sleep.

As to food, the mathematicians thought that Gulliver would also need 1728 times as much as a Lilliputian, each and every day, and they regarded doubtfully the efforts of a Lilliputian chef as he busied himself plucking a lark, in Lilliput smaller than one of our flies.

However, the mathematicians were at this point mistaken, because in the resting condition the bodily energy requirement increases in proportion to about the square of the linear dimension* and not in proportion to the cube, which of course does apply to the number of cells in the body. In the larger body, the "combustion" is a slower process and as a rule the bodily movements are also slower.

The reason for this is that mechanisms exist to maintain the body temperatures of "warm-blooded" creatures at levels at which the enzymes involved in metabolic changes function at their best.

In an active and lively mouse, the "engine" runs so fast that per gram of body weight, the mouse gives off 6 times as much heat per hour as man does. And from the fact that *Escherichia coli* exceeds humans in heat production by a factor of 400, we can conclude that this small creature is astoundingly active.

To those who are interested in making comparisons between the achievements of men and machines, the following may be instructive:

- The energy output of a man, walking, is, weight for weight, similar to that of an ocean liner.
- The energy output of a banana fly is, weight for weight, similar to that of a car driven fast.
- The humming bird's cruising speed is about 55 km (35 miles) per hour; for this, the "energy output/weight" ratio is higher than that of a modern helicopter flying at the same speed. The tiny creature would rapidly "burn itself up" if it did not go into hibernation during the hours of darkness, in which period its metabolic rate drops to one-fifteenth of the daytime value.
- For the bacterium *Escherichia coli*, the energy output is, weight for weight, similar to that of a jet aircraft in level flight.

An animal would encounter serious trouble if its metabolic rate were in some way forced above the normal, because of the "cube/square" law. The elephant, with its weight of 4000 kg, would need a skin area

* The exact figure is $d^{2.25}$

60 times that of a man if its metabolic rate were the same weight for weight, but in fact the elephant's skin is only 20 times the human measurement, so the elephant lacks $\frac{2}{3}$ of the needed "heat loss." If its metabolic rate were that of a mouse (again weight for weight) the elephant would die in a few minutes from "heat stroke."

8.18C Gluttony and Fasting

The intensive metabolism found in the tiniest creatures, which are best able to dispose of body heat, leads to a great intake of food, based on body weight (see Section 8.18B). The performance of very small living creatures seems to us men real gluttony. For instance at one "meal" a louse takes up 1 mg of its victim's blood, which, weight for weight, is as if a man's dinner amounted to 40 kg, more than half his own body weight.

The record appears to be held by the gnat, which at one bite (really one "prick") takes up its own weight of blood, corresponding to some 100 human meals, that is about equal to man's requirements for 1 month (70 kg = 150 lb).

It is thus understandable that the larger creatures make a better showing when we come to consider fasting; a crocodile can be satisfied for a half-year with an amount of food equal to its own body weight. Pythons have been known to go without food for 4 years, but the tiny shrew cannot live more than about 3 hours without nourishment.

We men come between these extremes; survival without food is possible for 60 days, if we can somehow find water.

8.19 Biological Bibliography [5]

An optimistic biologist tells us that only a mere 1000 papers on fleas have been published during the past 5 years concerning the 1500 known species of these animals. If the number of publications on beetles—of which 300,000 kinds are known—were to appear in the same proportion, there would be 200,000 beetle publications, more than the whole world literature existing . . . say before Columbus discovered America.

8.20 Vitamin D and Your Neighbors

Ultraviolet radiation from the sun (and sky), striking 1 cm^2 of an unpigmented—"white"—skin, is able to synthesize 36 International Units of vitamin D in 6 hours.

If the occupant of such a skin were to expose himself to a tropical sun in "birthday costume," his synthesis of vitamin D by the exposed part of his body would be at the rate of about 250,000 International Units in 6 hours. This would be enough for about 300 of his friends and neighbors, if he had any way of sharing the quantity of vitamin surplus to his own needs.

8.21 The Blood Bank

The worldwide average for the weight of all human beings, including ballerinas, strong men, and pygmies being about 65 kg (140 lb) the weight of the Earth's population of about 3500 million exceeds 2×10^{11} kg.

In the human 7% of the body weight is blood, so that the total amount "in circulation" in the world is about 1.5×10^{10} liters (0.3×10^{10} gallons). This is a comforting thought for our "blood-transfusion" services; yet the quantity could be contained in only 50 of the super-tankers now being developed, for each will carry a 300,000-ton cargo.

One liter of blood contains 5×10^{12} red corpuscles; on the Earth there are thus about 7×10^{22} altogether. But let us consider this multitude in a realistic way: it approaches the number of molecules in 1 cm^3 of water, 10^{23}!

Finally, each red corpuscle contains about 300 million hemoglobin molecules, so the total number of these on the Earth is just over 2×10^{31}. This is about the number of water molecules in the cargo of a 200-ton coaster.

8.22 Our Balance Sheet for Iron [44]

As to the iron within our bodies, we are very thrifty, but it is quite otherwise with our use of iron in general.

- The internal balance:

  In an average healthy human body there is about 4 grams of iron, the weight of 4 medium nails. The daily intake—and thus rejection—of this important element is about 1 mg, the weight of the point of one of these nails.

- The external balance:

  The worldwide average use of iron, referred to the whole population of the Earth, is $\frac{1}{2}$ kg per person per day, about the weight of the pair of pincers which we should need to pull one of the nails out of a piece of wood.

One-half of the iron found in the human body is in the form of hemoglobin, a complex molecule containing mainly carbon, nitrogen, oxygen, and hydrogen to the tune of about 1 iron atom among 1500 of the others. The Earth's main store of iron, apart from that in living matter, is of course iron ore with about 1 iron atom among 1.5 others: 1000 times as rich in iron as hemoglobin.

Hemoglobin, which is in fact the coloring matter of the blood's red corpuscles, is formed in the bone marrow, and, circulating with the blood, the molecules have an average life of 120 days. This is in sharp contrast to adenosine triphosphate (ATP) of which the molecules are estimated to be decomposed and re-formed 2000 times in a day (see Section 8.17). Still, even with the average 120 days' circulation of the hemoglobin molecules, there is—figuratively—feverish activity in the bone marrow to maintain the vast number, 2×10^{13}, of red corpuscles which should be present in the normal healthy man's 5 liters of blood. To do this requires the breakdown and re-formation of 2 million red corpuscles per second, a number approaching half the population of Denmark.

The products of the breakdown of the hemoglobin disappear from the body, with the important exception of the iron. This is fortunate, because the quantity of iron associated with the "discarded" material from hemoglobin is about 20 mg a day: 20 days would be needed at the body's normal rate of intake of iron, 1 mg per day, to replace the 1 day's loss.

If the body failed to enforce "No Exit" on the greater part of the iron involved in biochemical breakdown, we should have to begin to think about the 4 nails, the whole iron content of our bodies. However, we should have about half a year for this, with the iron disappearing at about 20 mg a day, ignoring at this time the relatively very small effect of our normal iron intake, about 1 mg a day.

Finally we may note that loss of about half the volume of our blood usually means death. However, in such a case there should be, somewhere or other (if we could only find it) the iron for 2 more nails in the coffin!

8.23 Watch the Temperature

As a general rule, the speed of a chemical reaction in an inorganic system is doubled by every 10°C rise in temperature.

For some organic materials, and in particular for proteins, this speeding up with rising temperature is very much more marked; a 1200-fold increase in speed for 10°C rise in temperature is no exception.

Because of this, it is fortunate that the outer layers of the skin have a strong retarding action on heat transmission to the inner and more sensitive layers. Nevertheless a severe burn—a "third-degree burn" in doctors' language—

which would result from 7 hours at 44°C

would also result:
- after 5 minutes at 50°C
- after 5 seconds at 60°C
- after 0.5 second at 70°C

8.24 Splitting Cells and Splitting Atoms

An up-to-date riddle might compare the natural proliferation of cells in the human body with the splitting of atoms in a "uranium" bomb. In man, growth involves the continual splitting of cells into two parts, each of which, nourished by food material, grows and, on reaching a certain size, divides again, and so the process goes on.

The number of such divisions which take place from the time of conception to birth of a fully developed child is 47, based on the assumption that the child's body comprises 10^{14} cells. This follows from the relation $2^n = 10^{14}$, in which $n = 47$.

So even after only 1 year from conception our body cells are of about the fiftieth generation.

In an atom bomb weighing 1 kg (2.2 lb), the number of successive splittings during the explosion is 61; however these 61 "generations" come and go in a small fraction of a second.

Perhaps our riddle should have been to ask "How is a man like a nuclear reactor?" for the bomb's "generations" are unimaginably more hurried than the cell splitting in the human embryo, albeit of the same order numerically.

8.25 Seen by Fish

When the sun rises just above or is just about to set below the horizon of the sea, its rays are bent to 49° from the vertical as they pass through the surface of the ocean. A fish's view of the whole scene above the surface of the sea is thus compressed within a cone of only 98°.

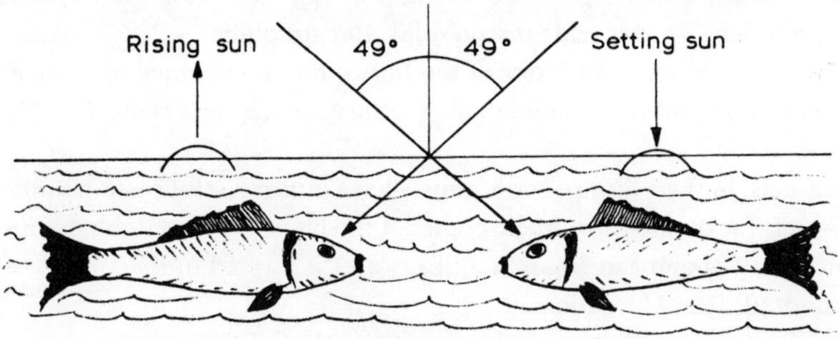

A fish's view compressed within a cone of 98° only.

. . . *and by Flies*

Dragonflies can be said to see as much, at a given moment, as the inhabitants of a small town—population 15,000—for their eyes have, combined, a total of 30,000 facets.

8.26 Lives Short and Long

We can take the present maximum lifetime which a man can expect as 120 years; calling this 1 minute, on a new time scale, the period of time since the solidification of the Earth (3.5×10^9 years ago) becomes 50 years.

The oldest material so far found which may be termed "living" is bacterial matter from Lower Cambrian salt beds which, on release from the mineral mass, resumes the normal life of this type of organism, even though it has been so long underground, and its total age can be represented by 10 years on our "compressed" time scale.

As to other long-lived organisms, Sequoia pines in California have lived for 1 "hour"—6000 real years—and in the animal world, turtles hold the crown. These slow-living reptiles may bridge a century or two, which ages up to 2.5 minutes on our scale, even more than the oldest human may expect.

8.27 Odd Bits

- In one year there are as many generations of banana flies as the generations of man since the Crusades, 700 years ago.

- A North African species of mice is capable of mating no less than 100 times in an hour.
- A queen bee can lay her own weight in eggs in one single day. If a hen produced on this scale it would lay 45 eggs daily.
- In the evolutionary history of the horse, the appearance of a recognizably different type of animal has, on the average, required 2.5 million years.
- Camels, in their hump—or humps—have a most useful reserve source of water, for by the use in their bodies of the fat reserve from the humps, which may amount to 50 kg (110 lb), water is formed in the amount of around 50 liters (11 gallons).
- White ants are so effective as excavators that, if they were scaled up to human size, they would have no difficulty in driving a tunnel of 100 meters diameter (about 330 feet).
- Germinating grain gives out heat, in terms of calories per gram at about 70% of a man's heat output of 1 calorie per gram (of body weight) per hour, thus at about the same rate as that for a snail.
- In a European dairy herd, the cows yield about 6 to 7 times their own body weight in milk each year—4000 liters, about 900 gallons.
- While a man is asleep, his heart, no larger than his fist, pumps about one teacupful (60 cm^3) of blood into his circulatory system per second. But if, after breakfast, he has to run to catch his morning train, the heart has to increase its output to a beer-bottleful (300 cm^3) per second.
- In the past year your red corpuscles have traveled a distance equal to $2\frac{1}{2}$ times round the world.
- In one year, you drink about 13 times your own weight of water, and breathe about 7 times your own weight of oxygen.
- Ant hills, with a population up to about that of Amsterdam in Holland, or Vancouver, Canada, about 1 million, surely are among the world's largest houses.
- Plankton in the Earth's seas and oceans takes up more of the element iron each year than the worldwide production of iron in blast furnaces.

Chapter 9

Man and Science

If by reading You can replace cocksure ignorance by thoughtful uncertainty, You have reached something.
(B. S. Henderson)

9.1 Speech Is Silver, Silence Is Gold

Men, even those most highly developed, often make only sparing use of the many possibilities open to them. This applies indeed to man's own creation, language.

It is excusable that a child 5 years old uses only 1%, or 1500 of the 150,000 "dictionary" words that a modern language contains. But this small treasury is used very intensively by him during the years of learning—the "silver age"—at the rate of about 1000 words an hour, so from 7000 to 15,000 words each day for an active child.

School quickly increases his vocabulary, but leaves him with less opportunity for using it; the gold of silence—attention—makes his silvery little voice dumb. At 10 years old, his treasury amounts to 7000 words of which about 30% are used actively (the remainder are used seldom or never) and in an hour he will use about 700 words.

On entrance to a university his word inventory will have grown to 20,000, and on graduation to 60,000, but of these only 10 to 20% will be in active use—a normal proportion in general.

A simple farm laborer, with his vocabulary of about 1500 words in active use, that is only about 1% of the possibilities of his mother tongue, has thus only just raised himself above the level of the 5-year-old child, or has fallen back to this state. In general the woman's treasury of words is noticeably less than that of her husband. Some have estimated the proportion at 1 to 4, although she is doubtlessly not less active in speaking!

For the average citizen the vocabulary is estimated to include 3000 words.

Naturally the literary experts are greater "language-consumers." It has been determined that an average American journalist uses but 6000 words in his newspaper articles, thus 4% of the available arsenal.

However, at a higher literary level, it has been determined that Victor Hugo used an active vocabulary of 38,000 words, Shakespeare,

24,000 words, and Homer 8500. But Horace, with 4600 words, came significantly below the American journalist, and Xenophon, with 3200, was about equal to the present-day "average citizen."

It is also of interest that the vocabulary figures for the Old Testament, with 5800 words, and the New Testament with 4800 words are less than the words in a present-day newspaper which uses about 6000.

9.2 The Human Brain and Its Brainchild: the Computer [45]

It is interesting to compare the performance and capacity of the human brain with the present capabilities of its most sophisticated brainchild: the modern computer.

- The first point of comparison should, perhaps, be the *degree of complexity*. This is a fundamental property, because a tenfold increase in it can convert a machine capable, say, only of *addition* and *multiplication* to one able to cope with *division*, *raising to powers*, and *interpolation*. In this respect, the brain is without doubt vastly superior, possessing more than 10,000 million cells, whereas a computer with "only" a million transistors already presents considerable problems to the electronic engineer. It is safe to say, then, that the brain of a single human being will long remain the equal in complexity of several thousand computers taken together.
- Under the heading of degree of complexity can also be included the concept of *memory*. Concerning human memory, estimates vary very widely as to its capacity for retaining information, some putting it as high as 60,000 items, each as complex as the multiplication table. By contrast, the storage capacity of a modern computer of 3 million bits* corresponds to a memory able to recall only 2000 such items, which means that the human memory may well surpass that of even an elaborate computer by a factor of a thousand or so.

It is astonishing to realize that if the human memory is indeed capable of retaining 60,000 complex items of information (10 million bits), it must acquire the bits at an average of one every 20 seconds!

– The human brain can receive impressions from the world outside it and make independent correlations between them. A man deprived

* In information theory one bit is the amount of information necessary to specify one of two equally possible alternatives.

of his five senses would be comparable to a computing machine with a fixed tape, for such machines can only *imitate*. Computers must be programmed by man; man's brain programs itself.

● In the matter of *economy of energy*, too, the human brain excels the computer. The entire brain, weighing about 3 lb, with its myriads of cells, operates on less than 100 watts, while a modern computer would consume the power output of a generating station designed to serve a community of 200,000 people, if it had to activate a like number of cells (transistors).

– On all the above counts, then, the computer is certainly no match for the human brain; but what are the compensatory advantages which have won it so important a place in the world of commerce, industry, and science and given it so vital a role in military defense planning?

Preeminent among its virtues is *speed*. The computer can operate a million times faster than the brain, for while the same nerves cannot be "triggered" more than about 100 times a second, the transistors of the computer can be switched on and off a hundred million times in the same period.

● When one considers also that computers need not suffer from fatigue, and will not normally make mistakes, their advantages are quite apparent. A few examples of their time-saving capacity will demonstrate how seemingly limitless their possibilities are.

– A fast computer could perform in roughly 10 minutes all the calculations which the average clerk would normally reckon to carry out in the course of his entire career!

– During the last century a Danish schoolmaster calculated the value of π to eight hundred decimal places—the feat occupied his whole life. A medium-size computer took a few hours to check his figures . . . they were free of error but it took the machine only 2 hours to calculate π to 1000 decimals.

– Computers of the style developed in 1950 could make calculations with numbers consisting of 12 ciphers, such as 875 426 319 764 × 183 915 632 284 at a rate of 3000 per second. That means one per second—

  • in 1950 for everybody in a small village (3000).

Then human brains developed their mechanical competitors so rapidly that the machines became able to make one such calculation per second:

- in 1960 for everybody in a provincial town such as Bound Brook, New Jersey (10,000)
- in 1963 for everybody in a city such as Cambridge, Massachusetts (100,000)
- in 1964 for everybody in a city such as Brussels (1 million)

and the forecast is, that computers will do this one calculation per second:

- in 1970 for everybody in New York or Greater London (10 million)
- in the near future for everybody in the free world . . . (1000 million)

The machine will then contain 10^{11} calculating cells which can be stored in a matchbox.

9.3 Some Linguistic Notes

● King Carel V of The Netherlands born in 1500 had a notable command of languages: Latin for his prayers, Spanish for his ministers, French for his Queen, Italian for his mistress, German for the knights, Flemish for his friends, and Hungarian for the Master of the Horse.

● Cardinal Mezzofanti, keeper of the Vatican Library (about 1800), is said to have mastered no less than 114 languages and 72 dialects.

If he had been invited to deliver a 1-minute address of welcome to an international conference in all the tongues at his command, he would have required over 3 hours to do so.

● But this is nothing, compared with the situation in India, which counts 700 ethnographical groups, each with their own language. Although the situation is less complex in Russia, tuition is given in 57 official languages.

● Though French is the mother tongue of only $2\frac{1}{2}\%$ of the world's population, its continuing use as an international language seems assured—no less than 35% of the delegates to the United Nations express themselves *en français*.

9.4 Recording Science in Bits [46]

It has been estimated that the total sum of our knowledge would require 10 billion (British usage, thus 10^{12}) words for its expression.

This means 3000 for each inhabitant of the Earth, or about the same as the number of words used by Xenophon (see Section 9.1).

In computer circles, binary digits are known as "bits," and assuming that 100 of these bits are needed to represent one word, 10^{15} binary memory cells would be needed to store our total compass of knowledge. Of this total compass, science and technology represent about 10%, or 10^{14} bits, and it is commonly supposed that of this, again 10% is that part which is really worth recording, thus 10^{13} bits.

If now we split up science and technology into 1000 sections, 10^{10} binary memory cells would be sufficient to record one such section. Seeing that the primary memory unit of a large computer may be expected to reach a capacity of 10^8 bits within a short time, and that secondary memory units are usually available to supplement the "primary," we can reasonably foresee that in the near future it will be feasible for a computer to store about 10% of all our scientific and technological knowledge.

Properly programmed, a computer can now carry out a "Literature Search" at the rate of 10,000 words per second, and run through a substantial branch of science within 3 hours. We can read at best, about 100 words per minute, and at this rate would need 10 years to work through the same amount of scientific material.

9.5 Recording Science in Books [47]

In 1880, it was possible to record all the important physical and chemical constants in one substantial volume of 250 pages, the famous "Landolt-Börnstein." It is now estimated that in the year 2050 a library of 4000 volumes, each of 250 pages will be needed for this purpose. The library will contain about the same number of pages as there are words in the Bible, that is about 1 million.

To give an idea of the amount of work needed to establish this sort of data, it is reported that the ("only") 600 pages of the current *Handbook of Biological Data* were based on the work of 4000 scientists whose own work was recorded on 20,000 pages.

Assuming the same proportion, our "2050 Library of Basic Data" will be based on the work of 7 million scientists—as many as the whole population of Paris—written on 35 million pages, that is 6 times as many as the numbers of letters in the Bible.

9.6 Recording Science in Periodicals

By the end of this century there may well be 25 million scientists and technologists active in the world, a number equal to about half the population of a large Western country, for example Great Britain or France.

- Their journals

At the present rate of increase the number of scientific journals will, by A.D. 2000, have passed 1 million: 7 times the number of "dictionary" words that a language of modern culture contains (150,000; see Section 9.1). There will then be 1 periodical for every 25 scientists.

- Their scientific publications

Every year more than a million scientific and technical papers are published, one for every citizen in a city the size of Birmingham, England, or Baltimore, Maryland, in at least 60 different languages: presumably many more tongues than are spoken in these two cities together.

The number of papers published up to now is estimated at, altogether, 10 million, corresponding to 1 for every group of 300 people in the world's population. But at the present rate of output, in 10 years' time each of our "groups" would have at its disposal 2 scientific papers to puzzle over.

- An encouraging outlook

For the physical sciences alone it has been estimated that if developments continue on the same scale over the next 150 years, the weight of the paper then needed "for the record" would be about equal to the weight of the whole Earth.

9.7 Recording in Daily Papers

The United States leads the world in the number of newspapers published daily, over 1700; it is followed by Japan, West Germany, and India, each with some 475 dailies. Similarly, and perhaps not surprisingly for anyone familiar with the gigantic Sunday editions seen in the United States that country leads by far in the consumption of newsprint —7 million tons of paper annually—followed by a mere 0.5 to 1.5 million tons in the other countries mentioned.

However if we compare the readership figures, the apparent lead of the United States disappears: first comes Great Britain with an average of 51 copies sold every year per 100 head of population; next, Japan with 42 copies per 100, and only third, the United States with 35 copies per 100 head of population.

9.8 Biographical Records [48]

The rate of growth of the number of entries in our biographical dictionaries is fairly constant, notwithstanding the exponential increase (that is, the ever-quickening growth) of the population as a whole. From this we must conclude that the greater proportion of persons worthy of mention in these dictionaries have already lived, and that the proportion of such in the population is falling.

However, every doubling of the population since the seventeenth century has been paralleled by an approximately eightfold increase in the number of scientists, so that the proportion of scientists in the world population has increased by a factor of 4.

It follows that at any time during this recent period, the majority of all scientists who, from the time of the early Greeks or even earlier savants, have ever existed on the Earth, were alive at the time considered, and this of course also applies to the present day.

Similarly, most of what is now known in the field of science has been discovered within living memory.

The upshot of all this is that there is much more of a past to live in and talk about if we discuss politics or wars than if we discuss science.

9.9 Science Invades Russia

Toward the end of the czarist regime in Russia, in 1915, the "army" of teachers in Russia, numbering 300,000, was smaller than the "Grand Army" of Napoleon which invaded Russia in 1812; but these few teachers were faced with a huge proportion of illiterate people, 75% of the population.

After the 1917 revolution another army of teachers, numbering now $2\frac{1}{2}$ million, 5 times the size of Napoleon's Grand Army, invaded their own country, and the opposing army of illiterates has by now been decimated, using this word in its correct sense of removing 9 out of 10 of the "enemy."

9.10 Development Is Being Speeded Up

Scientists tell us that—at least in their field—progress takes place so rapidly, that we may perhaps foresee a period characterized by the development of an invention in a year. In order to develop an invention into industrial exploitation it has been estimated [49] that it took:

• for photography	112 years (1727–1839)
• for telephones	56 years (1820–1876)
• for radio	15 years (1925–1940)
• for television	12 years (1922–1934)
• for the atomic bomb	6 years (1939–1945)
• for the transistor	5 years (1948–1953)
• for the integrated circuit	3 years (1958–1961)

9.11 A Gap between West and East [50]

9.11A The Ever Widening Gap

There are 800 million people in the rich countries, mainly Europe including Russia, and North America; their average per capita share in the gross national product*—to be denoted below as "income"—is some $1700 a year.

There are 2000 million people in the poor countries, which includes mainland China; their average income—on the same terms—is only $110 a year.

These figures mean that for each rich John West there are between 2 and 3 poor Abdul Easts, each earning $1 in the time that their prosperous friend earns $15. But country by country, the differences are still greater: in 1960 the average (North) American was 32 times, and the average Briton 16 times, as rich as the average Indian.

It seems promising when we learn that around the 1960's, because of foreign aid of the order of $10,000 million annually—a little over $10 contributed by the average John West—the total income of the Abduls increased at the same rate as the average income in the rich countries, that is at 4% per year. But unfortunately the real situation is not as good as this figure suggests, because the number of Abduls increases each year by 2.5%, against the Johns' increase of "only" 1.5%.

* In several Western countries this share in the gross national product is about 3 to 4 times the wages or salary actually earned.

Subtracting these increases from the income growth figure, 4% for each, we find that Abdul East earns 1.5% more each year, but John West 2.5% more. Consequently the relative gap is still widening at the rate of 1% a year, and, as is shown below, the gap is also increasing in absolute terms.

9.11B Closing The Gap

● *By Industrialization?*

In several of the developing countries the organizers of "five-year" or other long-term plans have had to accept the prospect of greater unemployment at the end of the plan-period than at its beginning. How is this paradoxical result possible?

In modern industry the capital invested is, on an average, $6000 per employee in the United Kingdom—about the price of a small house—against $3000 per employee in India—much more than the price of the up-country hut that the employee lived in before he started work at a factory.

India's third five-year plan called for the investment of $2800 million per year (amounting to 8% of the gross national product), a sum which would provide for nearly 1 million new jobs each year. But, because of India's high birthrate, the growth of the labor force is more than four times the number of new jobs in an average period. So, even if India were to invest all her available capital in modern manufacturing plants a considerable increase in unemployment could not be avoided.

From this situation, critical economists have deduced that it is necessary to develop to an intermediate stage in countries such as India, for instance "$300 per head" technology, instead of the existing figure for Indian "village" technology, estimated at $3.00 per head. This leads to another paradox, that large-scale labor-saving technology may be the least suitable means of development for Abdul East and his countrymen.

● *By Better Management?*

It has been calculated that if the managers of India's industries could achieve an output-to-capital ratio of 1 to 3, additional growth at the rate of 2% *could* result. Adding this figure to the annual 1.5% increase of the average Indian's income would bring his growth rate to the level of 3.5% per year . . . surpassing John West's rate of 2.5%.

This seems to be at least a beginning in closing the gap, but . . . the *absolute* gap would still become greater because—calculated now on the basis of 1965 figures—the Indian's 3.5% increase would mean only about $3.00 per year, whereas John West, with his increase of "only" 2.5%, would see his income rise by something like $60.

● *By Birth Control or by More Foreign Aid?*

From the foregoing it is obvious that a very effective way of arresting the increase in the "income gap," and perhaps even closing it, would be to achieve a reduction in Abdul's birthrate. This will need a very long-term educational program in a country of 700 languages and characterized by historical and other barriers against change.

Suppose that by a wave of a wand India could bring her annual population increase down from 2.5% to 1.5%—that is from 12 million (the population of The Netherlands) to 7 million (the population of Austria). It would pay to effect this reduction of 5 million (the population of Denmark) even at the cost of $1000 million spent on birth control, and this large sum is only 10% of the foreign aid now being received by India.

The other way of dealing with the income gap would be for John West to provide more aid than the 1% of the gross national product of the main donor countries, their 1963/64 contribution, which resulted in an addition of 6% to the average Abdul's "income," that is, his share of *his* gross national product.

One of the reasons why the effect of these gifts was limited is that in the period from 1957 to 1965 there was not only a fall in the prices of primary products, Abdul East's main exports, but also a considerable rise in the cost of the manufactured goods which he had to import. The resulting "trade gap" of about $10,000 million was of the same magnitude as the aid received from outside sources during these years, so that, from trading, Abdul was not much better off at the end of the period than he had been at the beginning.

It has been suggested that if the richer countries could spare not 1%, but 2%, of their gross national product for aid to the poorer countries, there would be a much better chance of the poorer economies "getting off the ground." Would this be too heavy a burden for John? Hardly, when we find out that the amount of money invested by the richer highly industrialized countries for military purposes alone is estimated to be about the same as the aggregated gross national product of the poorer countries.

To sum up, a contribution that John West would hardly notice could mean the difference between poverty and at least reasonable prosperity for Abdul East.

9.12 Gaps between West and West

Besides the ever-widening gap between West and East, another one is opening within the Western economies themselves, in fact between Bill Amerigo* from the United States and Karl Eureb† from Europe, as illustrated by the following:

● The scientific *brain-drain*—85,000 scientists, engineers, and physicians emigrated between 1950 and 1965 from Europe to the United States—is estimated as equaling an investment in education of the order of 100,000 million dollars. Put another way, this amounts to a gift by Karl Eureb of the fruits of 5 days of work per year in this period to Bill Amerigo.

● Whereas before World War II (1940) there was 1 *Nobel laureate* in the United States for every 10 European laureates, after the war the proportion had become 50:50. Here an existing gap is being closed.

● In 1960, of Bill's children over 16 years old, 11% received higher education, against only 3% of Karl's 16-year-olds.

● Although European science had its beginnings before Amerigo Vespucci lent his name to Bill's country—books were being printed in Europe 50 years before this event in 1507—the Bill Amerigos now spend $3\frac{1}{2}$% of their gross national product on research, as against the Karls mere $1\frac{1}{2}$%.

● Bill Amerigo earns 10 times as much from *patent licensing* and "know-how" arrangements as he pays out to others for knowledge of this kind, but Karl and Charles—West Germany and France—earn for themselves less than half of what they pay for such items.

● In the non-Communist world Bill Amerigo produces:

40% of the "old fashioned" products such as steel

* America's name is derived from Amerigo Vespucci, the discoverer of continental South America.

† The name Europe is derived from the Phoenician word Ereb or Eureb, which means darkness (setting of the sun).

60% of modern products such as bulk chemicals

70% of advanced products such as motor cars

85% of the "frontier" products such as aircraft and electronic equipment

● Although the Amerigos together number about the same as all the Eurebs, the first produce in the United States about twice as many goods and twice as much in services as the European totals of these.

Moreover, looking at *all* other countries, although the Amerigo community constitutes only 6% of the world's total population, their production of goods and services is no less than 30% of the world total.

9.13 Urbanization and Urbanization [26]

In all the industrial countries the number of people occupied in agriculture has decreased in absolute as well as relative terms.

● *In Europe*

In the year 1800, of the 190 million people then living in Europe, 2.5% (as it happens, the present population of Denmark, 5 million) were living in cities of 100,000 or more inhabitants.

Now, only a century and a half later, a truncated "Western" Europe—meaning west of the "Iron Curtain"—has a population not far from 300 million. Of these, 40% (120 million) are town or city dwellers, i.e., living in centers of at least 100,000 inhabitants, a number equal to the combined populations of Denmark, France, and West Germany, and 24 times the "1800" figure.

● *In the United States*

The net migration between 1920 and 1960 averaged about 700,000 per year, that is the present population of a city such as Boston, Massachusetts.

In 1964 only 7% of all Americans, that is something like the population of New York, were actually earning their living from agriculture; the comparable figure for 1880 was 44%.

9.14 New York, the Mammoth Metropolis [26, 52]

Among the world's metropolises—here defined as cities with over 1 million residents—New York, with some 14 million, is unique not only in the size of its population, but also in other respects. With "only" 9% of the country's population, it provides:

- some 40% of all jobs in nationwide wholesaling activities in the United States;
- more than 30% of the jobs in finance;
- almost 25% of the business and professional service jobs;
- and, somewhat surprisingly, almost 12% of the jobs in manufacturing.

Half of all these activities are within 5 miles (8 km) of Times Square in Manhattan, and every day the population of another metropolis (3 million) come to work in that part of the city.

In the crowded area of New York City the population density is 25,000 per square mile (10,000 per km^2). An apparent paradox is that the density drops as the population grows. The reason is that the territory covered by the urban agglomeration is increasing faster than the population: it grew indeed by 50% between 1950 and 1960, whereas the population growth was only 15% in this period.

9.15 Repeating the French Revolution?

Just prior to the French Revolution, the 25 million inhabitants of France were financially and economically dependent on about 250,000 nobles and clericals, i.e. effective power was in the hands of only 1% of the population. This was capitalism in an extreme form.

At present 85% of the world's capital is controlled by roughly 20% of its population, but due to the disproportionate rise in the birthrate of the "have-nots" it is estimated that by 1970 as much as 90% of this capital will be concentrated in the hands of only 8% of the population.

In the field of industry in the United States we find that, leaving out agriculture, but including transport undertakings and also the sometimes mysterious world of finance and banking, as much as two-thirds of the whole is controlled by only 500 separate undertakings.

We may take it that in most of these enterprises decisions of economic significance are made for the most part by various groups and committees, rather than by individual financial or other wizards or even tycoons.

Supposing that there are 100 persons in each of these groups, then the activities of two-thirds of the industries of the United States are dependent on the decisions of only 50,000 persons, less than 0.03% of the total population. Even assuming 1000 decision-makers for each enterprise—surely too high an average figure—the percentage of total population of the United States is only 0.3.

To give another impression of scale, it is estimated that the annual budget of the American Telegraph and Telephone Company, one of the largest private undertakings, would come after the first 12 states in order of size of their annual budgets; it must be added that this far exceeds the budgets of any of the Latin-American countries.

9.16 Incomes A.D. 2000

The present per capita share of the gross national product is estimated [49]:

• in the United States	$3500
• in Western Europe	$1800
• in U.S.S.R.	$1000
• in, for example, India	$100

It has been forecast [25] that rapidly increasing industrialization will result in a still more uneven distribution of annual earnings:

- In "postindustrial" countries, $12,000 (only 4 countries: United States, Japan, Canada, and Sweden).
- In highly industrialized countries such as Great Britain, West Germany, and Russia, $3000.
- In industrialized countries such as Mexico and Malaysia, $1000.
- In industrializing countries such as China and India, $300.
- In nonindustrializing countries such as Congo, $100.

So the "postindustrial" peoples will be at least 100 times as rich as the absolute "beginners," and some 40 times better off than those who have made a start toward full industrialization.

9.17 Twins, Triplets . . . X-Tuplets [53]

About 1 human pregnancy in 80 may result in the birth of sets of twins. For triplets the rate is about 1 in 80^2 and for quadruplets about 1 in 80^3 and so on. Perhaps fortunately, in this particular connection, not all pregnancies end in live births, and this is especially true of multiple births, of X-tuplets in fact, as "X" becomes larger.

Without this restrictive effect, the number of individuals born as twins over the years to the roughly 80,000 millions who have inhabited the Earth since the beginnings of recognizable "man" would have been 2000 million, a number about equal to $\frac{2}{3}$ of the Earth's present population.

The total number born as triplets would have been about 36 million, thus close to the number of individuals in Spain, and the total number born as quadruplets around 600,000, nearly the population of a town such as Manchester, England.

About 2000 sets of "quins" could have been expected, but imperfections in the records of less-developed countries, and the high antenatal death rate in such cases, have left us with only 5 or 6 definitely known sets.

Expectation of sextuplets is inherently very small, indeed theory would lead us to expect only 25 sets in our globe's 80,000 million inhabiters since man's beginnings, but we have a reliable record* of a septuplet birth, expectable but once in an aeon—1 chance in $3 \times 80{,}000$ million. This was in Hameln in Germany, in 1600, where it is still remembered by an inscription on the house where the sevenfold "happy event" took place.

The septuplet of Hameln.

There is one octuplet on record, born in 1921 in Mexico.

* There are many unrealistic records.

9.18 Our Hearts Beat Even Mammoth Tankers [24]

There seems little doubt that the human heart is by far the best pump that we know of. In a healthy man it will go on working for 75 years without interruption or irregularity, and with no repairs in the ordinary sense; every minute its 70 or so "beats" displace 5 liters (about $1\frac{1}{4}$ gallons) of blood. So, in an average lifetime the heart pumps enough blood to fill a 200,000 supertanker. And if we think of the heart of the Canadian Pierre Joubert, who is known with some certainty to have lived to the age of 113 years, the total volume moved was enough to fill a 300,000-ton vessel, a giant that is only now in active development, 150 years after Joubert's death.

9.19 Typewriter Language

There is a well-known sentence in use among typewriter repairmen to test the key action:

"The quick brown fox jumps over the lazy dog."

There are only 35 letters in these 9 simple words, yet these include all 26 letters of the alphabet, and form a rational and easily pronounced (and remembered) sentence.

However it must not be imagined that it is an easy matter to find an alternative sentence with the same merits. While it is true that there are 12 million possibilities of word-making with words of not more than 5 letters (as in our example) and using all the letters of the alphabet at least once, let the reader try to form these words into an intelligible and easily pronounceable sentence.

Although the restriction that "each letter of the alphabet should be used once and only once" leaves 8 million theoretical possibilities, it is still more difficult to find, among all these 8 million potential aphorisms, one which seems to make sense and which can be pronounced and memorized. Good luck to the seeker after truth in 26 letters.

9.20 America, the New Old World

In the state of Washington archaeologists have recently discovered material 12,000 years old which they attribute to the first man known to have inhabited North America, the so-called "Marmes" man.

Presumably he came from Sumeria (Mesopotamia) by way of the Bering Strait, which was at that time frozen over.

According to this dating, he had been in America 8000 years before the Sumerians developed their cuneiform writing.

For all that, Marmes man is a bit of a newcomer compared with the Oldoway man, who appeared $1\frac{3}{4}$ million years ago and is now presumed to be the first really manlike being on the Earth.

The interval between the periods in which these two early ancestors of ours lived was a very long time: for each year which has elapsed since "Marmes" man flourished, we must reckon on 150 years when thinking about Oldoway man.

9.21 Printing at the Limits

Recently readable text has been produced at reduction ratios of about 50,000 to 1 in area. Such microscopic printing approaches the optical limits of textual reproduction.

In the meantime, however, the possibility of etching on a thin metal film with an "electronic pencil" (80 Å in diameter) letters, 250 Å in height has been demonstrated.

But this is in principal not considered as being the ultimate limit. It should, at least theoretically, be possible to approach the level of coded chains as we find them in the DNA molecules (see Section 8.4).

The serious problem in the library will then not be storage, but retrieval.

9.22 Odd Bits

- At the time of Jesus Christ there were as many people in the world as there are now in Western Europe (300 million). The population is now increasing every day by as many persons as are living in Chicago.
- If the human race increases over the next 10,000 years as it has done over the last 10,000 years, there will then be a humanity filling a sphere the diameter of which will be 10,000 light-years (see, for meaning of light-year, Section 1.1).
- By the year 2000 the world's food supply will have to be increased by 4 times if an adequate (but not lavish) diet is to be available to everyone.

- During the past 50 years industrialized countries were able to cut the average working week by 30% due to mechanization. Continuing at this rate we would have to cut annually 15 minutes per week.
- More than half of the world's scientific literature is published in a language unintelligible to more than half the world's scientists.
- In 1805 only 5% of the population in England could read: a century later only 5% were analphabete.
- Each moment, all the year round, the number of people participating in international meetings is equal to the population of Summit, New Jersey, about 25,000.
- Not more than half the population of Europe enjoys "freedom of the press."
- The number of the possible combinations of all available technical data has been estimated, to be the fantastic figure, 2^{300}; for an idea of what this means, we mention that the number of atoms in and around the Earth is estimated at 10^{50}.
- Microfilm technique has progressed so far that it would be possible to record the whole of the Bible on a film the size of a postage stamp (2.5 cm^2, less than 0.5 square inch).

Chapter 10

Technical Flashes

The veritable discoverer is not the man who finds something new, but who first sees the significance of it.
(Max Verworn)

10.1 Vacuum and Pressure

If 1 cm^3 of air (containing 2.7×10^{19} molecules), were to be spread evenly over a surface equal to the area of the United States there would be about 250 molecules to every space the size of a fingernail (say 1 cm^2).

High-pressure techniques can achieve pressures of the order of 100,000 atmospheres. Even distribution over the area of the United States of 1 cm^3 of air under such compression would give each *fingernail* a population of 25 million molecules.

Modern ultrahigh-vacuum* techniques allow the number of molecules in a confined space to be reduced to only 3500 per cubic centimeter, which is only a thousand times the extremely low density found to obtain in interstellar space.

At one atmosphere (760 mm) air molecules collide with each surface atom every 10^{-8} second. This time increases in a more rarefied environment, such as that in space.

Consequently each surface atom of an aircraft is hit [54]:

- at an altitude of 200 km every 4 seconds
- at an altitude of 400 km every 100 seconds
- at an altitude of 600 km every 1000 seconds

And, every atom of a space vehicle, well under way to the moon is hit once every 6 weeks by an outside "air" atom.

At present the dimensions of ultrahigh-vacuum chambers range up to only a foot or two (60 cm), but larger ones are being designed to test the behavior of even complete aircraft.

As can be expected, it becomes extremely difficult to eliminate minute leaks of air into chambers of this kind. The designers of this advanced hardware regard as "intolerable" a leak permitting the entry of only 1 cm^3 of air per century. However small this quantity seems to be, the number of entering molecules then is still something like three times the world population per second!

10.2 Vacuums, or Vacua [55]

But even the high vacuum of 10^{-13} mm Hg mentioned in Section 10.1 is a high pressure compared with the conditions of certain chemical equilibria, provided that all other material could be excluded from the systems.

- Taking first limestone, lime, and carbon dioxide:

$$CaO + CO_2 \rightleftharpoons CaCO_3$$

At ordinary temperature the equilibrium concentration of CO_2 in this system is no higher than 1 molecule per 2.5 liters, which corresponds to 1/10,000 of the density of interstellar "space."

- Even this is high compared with our next example, for in the system cupric oxide, cuprous oxide, and oxygen:

$$2\,Cu_2O + O_2 \rightleftharpoons 4\,CuO$$

* Ultrahigh-vacuum = 10^{13} Torr = 10^{-13} mm of mercury (Hg).

the concentration of oxygen at ordinary temperature is 1 molecule in a volume as large as that of the Irish Sea, which corresponds to $1/10^{19}$ of the density of interstellar "space."

● To get close to an "absolute" vacuum, we must use the system cuprous oxide, copper, and oxygen:

$$4Cu + O_2 \rightleftharpoons 2Cu_2O$$

in which each oxygen molecule will have a volume 40,000 times that of the Earth to "play around."

Volume of the Irish Sea = 2700 km^3
1 cm^3 of air contains 2.7×10^{19} molecules
Volume of the Earth = 10^{12} km^3

10.3 Understanding "Traces" [5]

For chemists, the value and nature of some material may often depend less on what—to ordinary people—it is, but on what it is not: in other words what is in it that should not be there? In fact the proportion of impurities is often the factor which determines the quality or usefulness of the material.

To translate chemists' language about this subject into something easily grasped by "the ordinary man" we have constructed the first part of a dictionary, as below:

- 1% of impurities means 1 gram in a cup of tea (about 100 grams)
- 0.1% of impurities means 1 gram in a bottle (about 1 kg)
- 0.01% of impurities means 1 gram in a bucket (about 10 kg)
- 0.001% of impurities means 1 gram in a barrel (about 100 kg)
- 10^{-4}% of impurities means 1 gram in a ton (1000 kg)
- 10^{-5}% of impurities means 1 gram in a truckload (about 10 tons)
- 10^{-6}% of impurities means 1 gram in a barge-load (about 100 tons)
- 10^{-7}% of impurities means 1 gram in the cargo of a coaster (about 1000 tons)
- 10^{-8}% of impurities means 1 gram in the cargo of an ocean freighter (about 10,000 tons)

Impurities even on the minute scale last mentioned, that is at

0.000 000 01% can now be determined by use of the spectrophotometer, so that the classical problem of "finding the needle in the haystack" may be said to have been solved.

10.4 The Chemists' One-Step

Although the scientist of today has at his disposal means to attain temperatures ranging from one-millionth of a degree above "absolute zero" to 100 million degrees, yet the vast majority of chemical reactions are carried out in quite a narrow temperature range, namely from 300° to 1500° on the absolute, or Kelvin scale, corresponding to 27° to about 1250° Centigrade.

Representing the length of a thermometer capable of measuring the whole range of attainable temperatures—as above—as an hour's run in a car, or 100 km (60 miles), the mentioned range for chemical reactions would be but one long stride along the road, a distance of 1.2 meters, about 4 feet. And the difference of 1°C which on a clinical thermometer may mean, for a hospital patient, the difference between being "well" and "ill" will be represented by a distance of only 1 mm.

10.5 Ever More Light

Since Edison's invention of the electric (filament) lamp around 1890, in which a glowing carbon filament was the source of light, the following 65 years saw a gradual improvement in the efficiency of this sort of lamp from Edison's 8 lumens per watt to the 15 lumens per watt obtainable with the gas-filled metal-filament lamp, in common use in the home today.

But in the last 10 years there has been a much quicker advance from the 15 lumens per watt of the "home" lamp to 150 lumens per watt obtainable in the modern low-pressure sodium lamps used for street and similar lighting.

So, it would now be possible to "light" the whole state of New York with the electric power needed in Edison's time to provide lighting for New York City, and we may be sure that there will be still further progress in increasing this efficiency.

10.6 Photography at the Limits

In making films to portray life at its natural speed, the cameras are operated so as to take about 20 pictures per second (p.p.s.), and the films are projected at this same speed, but rates up to 4000 p.p.s. are now in common use in, for example, motor research.

Research groups active in plasma physics, aerodynamics, and related fields, require much higher speeds, however, and have actually attained something approaching 100 million p.p.s. Forgetting the practical difficulties of assembling such a multitude, a camera operating at this speed could make separate photographs of the whole world population, standing shoulder to shoulder, in roughly half a minute.

It is possible to use shutter speeds of such short duration (10^{10} p.p.s.), that each light quantum emitted by the subject is "photographed" only when it approaches to within 3 cm (1.2 inches) of the object lens. The quanta beyond this range are, as it were, "awaited" by the camera.

Perhaps the terminal limit in high-speed photography will be set by the time taken by light to traverse its own wavelength on its way from object to detector. This comes out at around 10^{15} p.p.s., and to illustrate the significance of this we may add that with this speed our contemporaries, assembled shoulder to shoulder as just described, would only have to stand still for about one-millionth of a second, during which time all would have been photographed.

10.7 Conversation by Lasers

The laser is a device which effects *l*ight *a*mplification by *s*timulated *e*mission of *r*adiation. By *pumping* (stimulating) the electrons in certain materials (crystals, gases, semiconductors, etc.) from lower to higher energy levels, and allowing the electrons at the latter levels to discharge themselves *coherently* (in unison), a very powerful and narrow beam of light can be produced. Such a beam, if directed at the moon, 385,000 km (240,000 miles) away, is so "narrow" that it would throw a circle of light with a diameter of "only" some 80 km (50 miles) on the lunar surface.

Provided suitable means of modulation can be found, it has been estimated that a single laser beam could be used to relay some 100 million simultaneous telephone calls or . . . all the radio and television programs being transmitted at any given time.

10.8 Something Concerning Watches

● *Gaining a Full Second*

The watch of the watcher on top of New York's Empire State Building (380 meters = 1250 feet above street level) tends to gain as a result of the decreased braking effect of the Earth's magnetic field. The increased tempo of the clockwork is, however, not too alarming ($5 \times 10^{-4}\%$): 70 million years would have to elapse before the watcher's watch had gained a full second.

● *The Atomic Watch*

By applying the maser (*m*icrowave *a*mplification by *s*timulated *e*mission of *r*adiation) principle, cesium crystals and hydrogen atoms can be made to produce resonance vibrations with an extremely high degree of constancy. Regulated by the tempo of these vibrations, an atomic watch based on hydrogen which started to operate when the first of the four ice ages which man has seen (and felt), 300,000 years ago, would now have lost no more than 1 second.

● *The Incorrect Watch*

Asked to judge between the reliability of a watch which lost 10 minutes a day and one which had lost its main spring, one may be inclined to decide in favor of the first. It is easy to calculate, however, that the losing watch would be correct only once in every 2 days, whereas the other would show the right time twice every 24 hours.

10.9 Quick Service

In one nanosecond—one thousand-millionth of a second—light (or an electric impulse) travels only 30 cm. So, in a computer the shortening of any circuit by 30 cm will save time to the extent of one nanosecond. But, not satisfied with this sort of time saving, experimenters have devised computer circuits in which the operative, or "switching," times are counted in picoseconds (1 million-millionth seconds); in fact, about 400 picoseconds, during which time light or an electric impulse will travel some 12 cm. Such a circuit would be about the size of an ordinary printed capital "H," and would contain 5 transistors, each of $\frac{1}{6}$ mm^2 area, that is $\frac{1}{3}$ of the cross-section of a human hair.

Moreover, we already have switches in telephone circuits not larger than the period (full-stop) at the end of this sentence which can make up or break off a connection in 10 nanoseconds, time for a car to run 3/1000 of a millimeter.

10.10 Porous Catalysts

One of the catalysts used in the process of hardening natural oils by hydrogenation is "nickel on kieselguhr," a porous material with a pore volume about 0.75 cm^3 per gram. The radius of the pores is, on the average, 25 Å.

Let us imagine all the pores in 1 gram of the carrier (kieselguhr) as cylinders of this same radius, 25 Å, and all these cylinders to be laid end to end.

The resulting tube would be nearly 40 million kilometers long, and a jet-fighter aircraft flying at 1000 km per hour would take $4\frac{1}{2}$ years to fly from one end of this long "pore" to the other.

10.11 Fibers, Other Than for Textile

- Crowded Paper

There are some 20 million fibers in a quarto sheet of good-quality writing paper, a number which represents more than twice the present population of Greater London.

- Desalinators

Twenty hollow fibers, each thinner than a human hair, making a bundle 30 cm in diameter and more than 2 meters long, are the elements in a modern permeator for desalination of brackish water, providing a permeation surface larger than the area of a football pitch. This number of "hairs" is what we normally find on 150 heads of white men or on 200 heads of black men.

10.12 Pure Materials

Uranium for use in atomic reactors should ideally be refined to such a degree of purity that not more than one borium or cadmium atom remains for every 10 million uranium atoms, corresponding, say, to a ratio of one red-haired person to the entire population of New York City.

Certain types of transistor require an even stricter standard of purity, the admissible inhomogeneities in the finished product being comparable to a situation in which only three red-haired people were represented in the world's total population, i.e., 3:3500 million.

10.13 Moving Buildings

Observatories are equipped with extremely delicate measuring instruments . . . with which some remarkable observations have incidentally been made.

At one establishment it was discovered that the observatory itself inclined 1/10 mm toward the East in the morning and leaned toward the West to a similar degree in the evening. From this it was apparent that large buildings in general follow the course of the sun to a measurable, if extremely limited, extent.

Other observations have disclosed that the observatories of Leipzig and Rome have moved a finger's length closer together over a period of 60 years, while in the course of 80 years the observatory at Heidelberg has become a hand's length more remote from the North Pole.

10.14 Falling

- *The Maximum Speed*

An object falling from any height is subject to the Earth's gravitational force, and therefore tends to accelerate. But as the object nears the Earth's surface air resistance comes into play, which, increasing more than proportionally to the speed (of fall), causes an end to the acceleration and the establishment of a fairly constant "terminal" velocity.

For a human being, this terminal velocity after a fall from a great height is about 240 km per hour (150 miles per hour) or about twice the speed of a rather fast car.

- *Falling from Mount Everest*

A helicopter pilot who intended to land on Mount Everest (29,000 feet = 8888 meters) but mistook the characteristic windblown plume of snow at the summit for the firm foothold of the pinnacle, would have three-quarters of a minute to reflect on his error of judgment as he plummeted down to sea level.

● *Heating Water by Free Fall*

It is estimated that water falling from the height of the Chrysler Building* in New York would be warmed by 1°C (1.8°F) on impact on a level surface—we disregard heat loss by evaporation during the fall. Supposing that free fall without dispersion was possible, we have at our disposal an ingenious, though perhaps impracticable, method of boiling water. We may let it fall from a height three times that of Mount Everest, itself 8,888 meters (29,000 feet) high.

10.15 An Accident with a Centrifuge [39]

In a modern ultracentrifuge the rotor, weighing about 1 kg ($2\frac{1}{4}$ lb) turns at, say, 50,000 revolutions per minute. If this rotation were to be instantaneously transformed into linear movement, the translational energy thus becoming available would be enough to stop a train of 5 (runaway) railway cars, each weighing 20 tons gross, approaching at 100 miles an hour.

10.16 The Engineer's Disappointment

Mr. Turner, an engineer, admiring under a microscope the surface of the turned and ground axles and bearings he was responsible for, was happy to state that the surface asperites were no more than a few microns (about 5μ) high.

His customer, not so easily satisfied, however, pointed out that if the diameter of a turned 1-cm (0.4-inch) axle, of which there are many in his car, were to be enlarged to the diameter of the Earth, the turning of such an axle in its bearing would be similar to an encounter of the Alps and the Andes (each about 8000 meters = 27,000 feet high).

What the customer desired in his car were axles with lapped specular surfaces (asperites 0.025 μ = 250 Å). Then the encounter would be reduced to a scale comparable with rural church towers, 30 meters

* About 300 meters or approximately 1,000 feet.

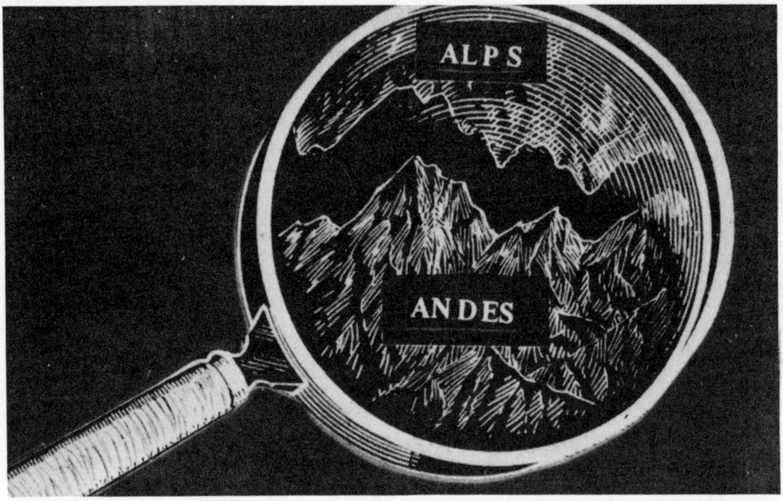

. . . the turning would suggest an encounter of the Alps and the Andes.

(100 feet) high. This seemed to him a better performance, although he was still afraid that some disaster might happen with towers scraping over each other.

Learning about all this, Mr. Turner calculated that filling up the empty space between the asperites of a turned axle with lubricating oil works out to a layer 10,000 molecules thick: for a lapped axle this would be 25 or 50 molecules thick. Then he began to appreciate that salesmen of lubricating oils are more interested in machines with turned axles than in those with lapped axles.

10.17 Improvements in Construction Technique

When it became possible to rebuild Holland's great bridges, destroyed during World War II, it was found that the same traffic capacity and permanence could be attained using only 50% of the materials used in the original constructions.

Supposing that the "Empire State" building, in New York City, with its 102 stories, were now to be built again, it would be possible to add 15 more stories without using any more steel than was used in the original building.

10.18 Ductile Metals on Ocean Bottom

A crack produced through stressing rigid materials like steel or glass progresses with a speed of about 2 km per second, roughly seven times the speed of sound (in air). If the Earth's crust possessed the characteristics of such materials, a crack would travel round its entire circumference within 6 hours.

In order to separate the atoms at the critical spots—the ends of the incipient cracks—not only the interatomic forces involved have to be overcome, but also any extra compressive force exerted on the metal. Consequently external pressure may result in raising the critical cracking stress [69]: actually zinc was found to be ductile at room temperature under a pressure of 1300 atmospheres.

Such pressures are found at a depth of 3700 meters (6700 fathoms) which is being called the "average depth" of the oceans. But locally depths more than 10,000 meters have been found, a promising outlook for deep sea traffic.

10.19 Progress in Welding

- Father welder, armed with an oxyacetylene torch, can produce a heat concentration of 10,000 watts per square centimeter.
- His son, the electric welder, can achieve ten times this value.
- The grandson, making use of the electron beam, is able to surpass his grandfather's performance by about one hundred times.
- Finally, in the hands of the great-grandson, the laser (see Section 10.7) is capable of developing temperatures far above the boiling point of any element, producing a heat concentration of the order of 1000 million watts per square centimeter.

10.20 Feeding on Alcohol

Considering the fact that the nutritional value of (ethyl) alcohol is about 9500 kcal per liter, it appears that in many countries a considerable part of the calorie intake of the population is derived from alcoholic beverages.

The French are in the lead: with an average consumption, in terms of pure alcohol, of 20 liters (4.4 gallons) per head per year, they cover the calorie needs of a man doing light manual work—typical for the average man of A.D. 1968—for 67 days, not far short of 3 calendar months, and more than this on the basis of a "5-day week"! The Italians come next, with consumption sufficient to cover nearly 2 (calendar) months on the same basis.

Next follow countries with, perhaps, a "Teutonic" tradition: Germany, Switzerland, and Austria, who consume enough alcohol to supply their calorie needs for something like $1\frac{1}{2}$ months on the calendar.

In the United States, Britain, and Canada the average suggests a more modest consumer, taking an amount of alcohol that would provide energy for only 20 days of light work, hardly a calendar month with even a 5-day week. This may reflect something in the Anglo-Saxon character or it may prove that "On the Rocks" may not seem to be the most animating recipe.

10.21 Rockets and Hoses

Rocket propulsion is an application of the simple mechanical principle that every force, or "action," must be balanced by an opposite force, or "reaction." In the garden we sometimes demonstrate this principle by turning on the water into a garden hose at full pressure: the end of the hose will move back from the direction of the jet, and sometimes it will whip to and fro like an angry snake.

To give an idea of scale, the initial thrust of a "V-2" type rocket—25,000 kg, or 56,000 lb—would be equaled by the assembly of about 28,000 week-end gardeners, each with their hoses turned on "full." In the event of man-power shortage, we could substitute about 550 members of the fire brigade, each again, with his hose "full on."

V-2 rocket, thrust = 56,000 lb
Garden hose, thrust = 2 lb
Fire hose, thrust = 100 lb

10.22 Invisible Wear

A large truck consigned to the scrap heap may have lost no more than 0.1% of its original weight and yet have been made unserviceable by this loss, through chemical or mechanical action, of only a pound or two.

A similar, apparently insignificant, degree of wear gradually frays the collar and cuffs of your shirt and causes you to discard a garment which is otherwise, perhaps, as good as new.

10.23 A Small Clearance

In modern automobile engines, the clearance allowed between piston and cylinder wall is 5/10,000 of an inch (about 1/100 mm); this is less than the thickness of the ink in the printed letters on this page.

10.24 "Thick" May Be Thinner than You Think

A friend of mine [24], impressed by the weight of a 3-cm ($1\frac{1}{4}$-inch) steel plate which he saw suspended on a shipyard crane where a 200,000 tanker was in course of construction, reconsidered his conclusion when, on his return home, his wife offered him a biscuit out of the tin. Being quick at arithmetic, he calculated that if the supertanker were reduced in scale to the size of the biscuit tin, that is, a $400\times$ "diminution," the thickness—0.28 mm (about $\frac{1}{64}$ inch)—of the tin would be 4 times that of the steel skin of the "reduced" supertanker.

Then, lighting up a cigarette (after noting that the thickness of the paper was 0.03 mm, or 0.001 inch) he came to the startling conclusion that if the supertanker were reduced still further in scale to the size of the cigarette (diameter, 8 mm), the cigarette paper would be 6 times as thick as the skin of the supertanker.

10.25 Cars on a Whisker

A car weighing 1500 kg (about $1\frac{1}{2}$ tons), could be safely suspended on a bundle of 50 steel piano wires each of 1 mm^2 (0.0016 square inch) cross section. New sorts of steel have now been developed that make it possible to reduce the required number of wires to 5. And even specially prepared sorts of glass fibers (etched and lacquered) exist today of which 10 could do the job, whereas it should theoretically be possible to hang the car on only one of them.

New developments in single crystal filaments ("whiskers" of tungsten) have produced materials of such extraordinary strength that on one 1 mm^2 wire it would well be possible to suspend two cars.

Tensile strength of ordinary steel 30 kg per square millimeter.
Tensile strength of tungsten whiskers 2500 kg per square millimeter.

À Propos
Another promising aspect of some of these new high-strength materials is their light weight. A racing car with a body of 30 kg (66 lb), half the weight of the driver himself, won in 1968 the Grand Prix of France. It was composed of carbon filaments: if steel had been used, the car body would have weighed as much as 7 drivers.

10.26 Rail Traffic Is Not So Simple

Rail traffic is more difficult to schedule than most of us think. For example, the number of ways in which 40 trains can be fed over a single junction in double track is 10^{18}. This is a million times more than the number of celestial bodies in our Galaxy.

10.27 Odd Bits

- Modern tele-objectives make it possible to measure an object at 3.5 km (4000 yards) distance, as if it were only taken 45 meters (50 yards) away.
- We can now make sharp photographs of earthy objects from aircraft, flying at sonic speeds at an altitude of only 45 meters (50 feet).
- Some 2.5 million chemical compounds are now known—nearly one sort for every 1000 persons—and an increase of 10% per year is reported.
- Some 500 different chemical compounds have been found in cigarette smoke and tars, many of them cancer-inducing or toxic when tested on animals.
- The boilers in a large electricity generating station, of capacity say 500 MW will include about 100 km (60 miles) of piping in the steam-generating and superheating sections as well as in air pre-heaters; this is about the distance from Washington, D.C. to Baltimore, Maryland, or from London to Brighton or Eastbourne.
- Polypropylene is one of the plastics as strong as steel, as soft feeling as silk, and weighing less than water.
- One cubic centimeter of a good quality of carbon black in rubber (used as reinforcing filler) with a particle diameter of 10^{-6} cm has a

surface area equivalent to that of a little house (300 square meters).

- The total length of the fire hose sold annually in the United States would reach from New York to Buenos Aires (8000 km or 5000 miles).
- In a good sized window-glass factory, the daily production per man employed is enough to provide the windows for two ordinary houses.
- A conventional train running on steel rails needs about 1 kg of thrust for each ton of its weight to overcome rolling friction at normal speed. The weight of a passenger including his clothes and his luggage, estimated at 100 kg, will add to the total thrust to be produced by the locomotive something like the weight of his hat (100 grams).

Chapter 11

The Hydroworld

Water was considered by the ancient Greeks as one of the four elements.

The falling drop.

11.1 Our Contact with Water [56]

Although there are 1.3×10^9 km^3 of water on or near the surface of the Earth, many of us have contact with only a very small proportion of this useful commodity.

● If all this water were evenly distributed over the Earth's surface, it would form a layer 2.5 km (1.6 miles) thick; 97% of all this water is in the sea. It serves only the sailor and some enthusiastic bathers. Of the remainder, two-thirds, or 2% of all water, is frozen in glaciers and in the polar ice caps, so only mountaineers, explorers, and a few other scientists are likely to come into contact with it.

● If this frozen water were melted and evenly distributed over the Earth, the film thus formed would be 50 meters (55 yards) thick. This melting disaster would result in the overflooding of most of our fertile grounds. One percent of the total water is all that is left for direct use in home and factory (1.3×10^7 km^3).

● If evenly spread over the Earth, this "usable" water would form a layer of 25 meters (27.5 yards). However, as all but 3% of this water is "ground" water, only geologists (and water-diviners?) have much contact with the bulk of it. We make the most use of flowing water, which is only 1% of the 3% of fresh water which is in fact above ground in lakes and rivers (4000 km^3).

● This flowing water, spread again as described above, would form a film nearly 1 cm thick. This thin layer disappears so quickly (streaming, evaporation) that it would only suffice for human needs for 2 weeks, but our stocks of "visible" flowing water are being made up all the time by rain and from underground sources—springs and seepages.

One more accumulation of water that we all know about is the water vapor in the Earth's atmosphere, for it is a major factor in the phenomena which we call "weather." Though it is only one hundred-thousandth of the grand total of all water, it amounts—on an average—to 13,000 km^3.

● This atmospheric water, condensed and evenly spread out would cover the earth with a layer 2 fingers thick (2.5 cm = 1 inch).

Concluding one can say that as a matter of fact out of the 2.6-km water layer available, our daily living conditions are defined by:

– the 1 cm-thick layer of flowing water
– the 2.5-cm thick layer of atmospheric water.

For our daily consumption the flowing water is most important but, although its quantity is small compared with the total water, it is still abundantly available to man. We drink, in a year, the same portion of it as if we were to read one printed character out of all the characters in a stack of all the 20-page newspapers of a local community journal printed in 20,000 copies per day, during a full year (ratio $1:4\times10^{12}$).

Volume of water on the earth	1.3×10^{9} km^3
"Flowing" water	4×10^{3} km^3
Water in the atmosphere	1.3×10^{4} km^3

11.2 The Rainmaker's Job [56]

The average rainfall around New York City or London, thus in climates that are, on the whole, though not always genial, amounts to 760 mm per year. It so happens that this is equal to the average annual rainfall over the whole surface of the Earth, and, again quite by chance, to the height of the mercury column in a common barometer.

Supposing that this rainfall was indeed equally distributed in time, and over the Earth's whole surface no rainmaker would have had any call on his services. However, at the town of Calama, in northern Chile, no one is known to have seen even a drop of rain. In contrast to this situation, the rainfall recorded at Mt. Waialeale in the Hawaiian Islands is 16 times the average amount in New York City. Here again we can notice a singular chance—the rainfall in a year at Mt. Waialeale is about the same as the column of a "water" barometer.

On the smaller scale there is also call for the rainmaker, or at least the "rain-regulator." In a sharp "cloud burst" it is possible for rain to fall so fast that in an hour the annual average for that same spot might be equaled.

Therefore it is surely not surprising that among primitive people, rainmakers were, and here and there still are, held in high honor.

In terms of physical science, water is soon evaporated, but it has a remarkable resistance to resuming the liquid form by the reverse process of condensation.

In "clean" air—that is, air free of smoke and dust particles—a degree of supersaturation with water vapor amounting to some hundreds percent can occur, so that the rainmaker's first task is to distribute condensation nuclei, for example a finely powdered salt, in the air, around which "cloud" droplets can form. But for this purpose he will need to inject into each cubic meter of air a number of nuclei equal to the number of citizens of the United States, about 200 million.

The rate of fall of the tiny droplets first formed is far too slow to produce the effect of rainfall of any consequence, and the rainmaker's second task is to make these droplets grow 1 million times larger, thus to congregate into armies, each of which will include more droplets than the number of men in Napoleon's Army as he fought the Battle of Waterloo—now in every cubic meter of air, we shall find about 200 of these "armies."

To increase the amount of rainfall the experienced rainmaker has recourse to ice crystals. Crystallization of water vapor as ice is most efficiently induced by the injection of microscopic clay particles into supersaturated air, since there is quite a close correspondence in shape and structure between the crystals of ice and of most "clay" type materials. Thus crystallization as ice follows quite rapidly when supersaturated air is "seeded" with clay particles.

This done, the rainmaker can take a rest, because the resulting "snowflakes" are heavy enough to fall and may well turn to large liquid drops.

Considering rainfall throughout the world, the quantities involved are really large—about twice the volume of the Red Sea falls each year as rain, a total of 420,000 km³. This is about 30 times the amount of water

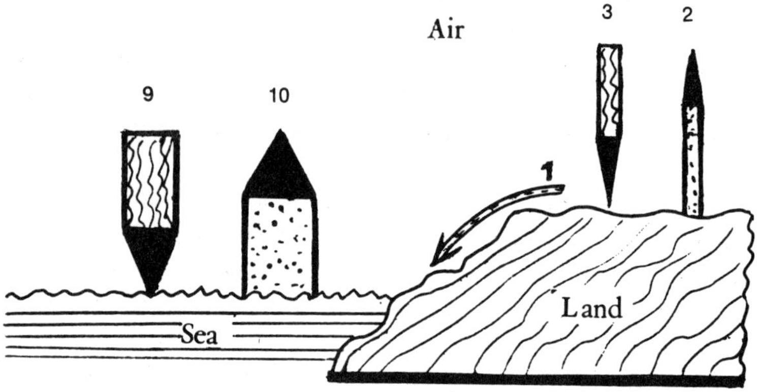

Interspheric water cycles. One unit corresponds to 1 million cubic meters per second, or $\frac{1}{3}$ liter per person per second.

held in the Earth's atmosphere at any given time, and it would provide a continuous shower bath for every inhabitant of the Earth, at the rate of nearly 4 liters per second.

The rainmaker of the past has never been aware of the following. If 10 liters of water evaporate from the sea, 2 liters do so from the land. Of these 12 liters, 9 come back into the sea and 3 on the land. Without the surplus of 1 liter on land there would be no runoff and we would have no rivers.

11.3 Within a Shower of Rain

In a heavy shower of rain, the actual raindrops occupy less than 1/100,000 of the volume of the air from which they were formed. In the picture the large square represents 1 cm³ of air; the little black square represents the volume of water formed by condensation of all the water contained in the cubic centimeter. If, however, there are 100 condensation nuclei, N, the amount of water is distributed over 100 smaller droplets.

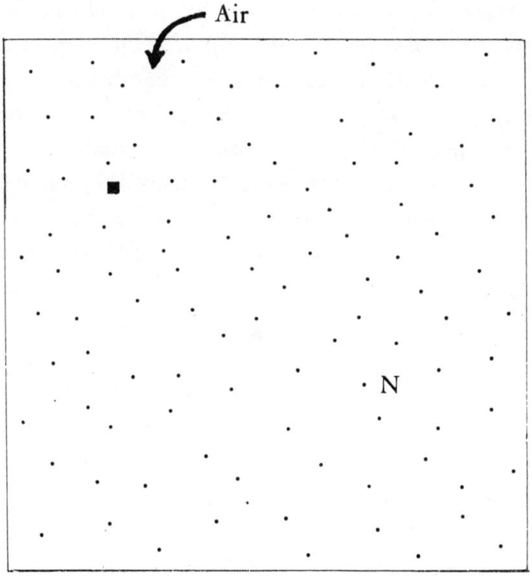

The amount of water, ■, formed from a cubic centimeter of air: 1/100,000.

It is as if in a good-sized room, of volume 100 cubic meters (3500 cubic feet), we were to find a whisky bottle ($\frac{3}{4}$ liter) filled to $\frac{1}{3}$ of its capacity with water.

11.4 Water as a Heat Regulator

11.4A The Human Calorifier

A party of scientists on a desert exploration suffered from the well-known troubles of the desert, unbearable heat by day, but temperatures below freezing at night.

One of the members, Professor Q'al-Orry, the distinguished physicist, remembered the market-gardeners' old trick of leaving tubs of water around in their fruit stores. When things are getting colder, this water loses 80 kilocalories (kcal) per liter as it freezes to ice, which of course slows down the cooling of the other contents of the store.

Starting from this recollection, the Professor considered the possibility of utilizing a human body as a heat engine, or heat buffer, on the same general lines. He recalled further that the body contains on the average 50 liters of water which on freezing would lose 4000 kcal, a quantity of heat which is about the calorie equivalent of the food intake of someone engaged in manual work, as for example a bootblack.

It is of course true that this food intake supports other activities besides cleaning shoes; it enables normal bodily activities to be carried on and also should ensure that our bootblack does not freeze in cold weather; but if this does happen he will have released about as many calories at the moment of freezing as he has just taken in as food.

Professor Q'al-Orry decided that the best way of utilizing a human heat engine, or "heat buffer," for the greater comfort of desert camps would be to let the "performer" eat as usual, but to expose himself to the desert's cold at night—within the tent—and to be remelted each day by the obvious source of extra heat, desert sunshine, in the daytime.

A human body, fully exposed to the sun's radiation, will take up heat at 250 kcal per hour, so that only to melt the ice in the human heat buffer, 16 hours would be occupied in supplying the 4000 kcal which we have seen above to be necessary. Sixteen hours is less than the period of daylight in the tropics; moreover it is doubtful whether a man's skin could withstand the rate of radiation suggested and certain that there would be appreciable heat loss by perspiration. However,

Professor Q'al-Orry, always optimistic, expected that conduction of heat from the hot desert air would make up the deficiency in heat without undue damage to the human heat engine.

Since the 4000 kcal heat of crystallization (or of fusion) of the water content of the human body is enough to warm an air volume of 13,000 cubic meters—that is, the volume of a sizable block of flats—by 1°C, it really depends only on the degree of insulation of the tent how many expedition members can benefit—and for how long—from the slow freezing of their colleague chosen as heat engine (or "buffer") and therefore necessarily unclothed, while the rest stay well wrapped up in their blankets.

Professor Q'al-Orry had almost persuaded one of the members to let himself be frozen by way of trial, in the expectation that the tent would not get colder than −1°C. But the potential "heat engine" was the equally eminent biologist John Ginnip-Igg, who, being well accustomed to the use of laboratory animals, proposed in spite of his own appropriate name, that the guinea pig for this trial should be the expedition's mascot, a lively fox terrier.

However this canine friend was too much of a favorite for any chances to be taken with his well-being; and the idea of a partly frozen camel as near neighbor in the sleeping tent being also unacceptable to the expedition, Q'al-Orry's scheme had to be given up.

11.4B Warmth from Cold Water

As expected, it became rather cold in the desert camp as night drew on, and Professor Q'al-Orry found himself awake, reflecting on some of the remarkable properties of water. He recalled several which have an important influence on warmth and cold on the Earth; some figures from his notebook—rounded-off—were these:

Specific heat of ice in the range −20°C to 0°C	0.5 kcal per liter per °C
Heat of fusion (or crystallization) of ice (0°C)	80 kcal per liter
Specific heat of water in the range 0°C to 100°C	1 kcal per liter per °C
Heat of evaporation (or condensation) of water (100°C)	540 kcal per liter
Specific heat of steam (water vapor) in	

the range 100–120°C at constant pressure	0.5 kcal per liter of liquid water per °C

Conditions on the Earth are greatly influenced by the relatively large amounts of heat evolved in or required for the two important "changes of state": water to ice and vice versa, and water to steam (water vapor) and vice versa. The heat which would be released by the freezing of all the 10^9 km^3 of water on the Earth once every year would be enough to warm up all the air in the atmosphere from -20°C to 0°C, ten times every hour. This suggested to Q'al-Orry, as he shivered in his camp bed, that the textbooks with their frequent references to the beneficial influence of the "warm" Gulf Stream (15°C, as it takes its northeasterly course across the Atlantic and along the coast of Europe) were wide of the mark since the heat evolved as the water cools from 15°C to 0°C is but a small fraction of that evolved when the water freezes, in fact 15/80 in round figures.

Apart from the minor influence of its salt content, the freezing of the Gulf Stream water, estimated at 10^6 km^3 per year, would release enough heat in the one year to warm the polar atmosphere from -20°C to 0°C not just once, but once every day for a year.

Glancing at his notebook again in the light of the desert moon, the Professor saw that the heat of evaporation (or condensation) of water was much greater—about 6.5 times—than its heat of fusion (or crystallization). After a little thought, he concluded that if the heat of condensation of the annual rainfall on the Earth (2.3×10^{20} kcal per year, based on estimated rainfall 4.2×10^5 km^3 per year) could be suitably applied, it would suffice to raise the temperature of all the ice and snow at present in glaciers and on the polar caps by 23°C.

Q'al-Orry had been a great reader of Jules Verne in his younger days, and now realized a possible theme for two more stories, perhaps as exciting as the 101 tales from that famous pen which we already enjoy. One was to be *The Giant Pump*, developing the idea of an enormous pumping station to be built to increase the flow of the existing Gulf Stream, and to establish a new "warm" stream directed toward New York and the eastern United States to save these regions—on about the latitude range of the Mediterranean Sea—from the severe winters which are their present lot. Doubtless Jules Verne would have placed the intake of his pumping stations where the water had already cooled to just above 0°C, so that the water would really "pull its weight" calorifically; in other words, to utilize the high heat of crystallization rather than the relatively much smaller sensible heat of cooling from 15°C to 0°C.

The other book was to be entitled *The Giant Bellows*. This would have described apparatus to transfer moist air from above the world's oceans to the polar regions, where release of its heat of condensation would then lead to entirely different conditions for the life of Eskimos and of anyone else in those hitherto chilly wastes.

A modern Jules Verne might tell us that the heat of condensation imparted to the atmosphere by the formation of liquid raindrops from water vapor in a really heavy downpour over the whole of Regent's Park, London, or one-half of Central Park, New York—in figures, 2 inches or 50 mm of rain over an area of 1 km^2—is about the same as the energy released in the explosion of the atomic bomb over Hiroshima in 1945 (see Section 12.6 where this is expressed in a somewhat different way).

11.4C The Energy Balance in the Arctic

The movements of warm air and warm water into the Arctic from the region immediately south are indeed a reality, though not of the scale envisaged by Jules Verne in his so far unwritten science fiction masterpieces *The Giant Pump* and *The Giant Bellows*. As a matter of fact the heat flows

- 150 kcal per square meter of the Arctic basin in the moving water
- 100 kcal per square meter in the moving air.

After his return from the hot and dry desert expedition, Professor Q'al-Orry plunged into the other extreme, a cold Arctic Sea excursion. He noted with interest that certain Russian technologists were considering a means of moderating the icy conditions along their northern coast quite different from Jules Verne's unpublished ideas. Their technique was based on the much higher absorption of radiant heat by water as compared with ice.

Now, in overall figures, the Arctic region receives—per square meter—about 600 kcal per year from the sun's radiation: the ice absorbs only 35% of this, about 200 kcal. But, if the ice were once melted, then the same water surface would absorb 85% of the heat energy available in the sun's radiation, 2.5 times better than things are now. For the Northern Hemisphere as a whole, this would mean an increase of 2% in the intake of heat from the sun. It seems sure that if the ice were once melted, absorption of heat from the sun would then be sufficient to keep the Arctic Ocean free of ice.

The Russian proposal is to spread a dark powder over the ice to enhance the absorption of heat; but where is so much soot to be found in these smokeless days? To make use of a simpler principle, one might prevent heat loss from the (relatively) warm water through evaporation by blanketing it with a monomolecular layer or one might supply heat based on nuclear energy to melt the ice. But perhaps it is as well for us and for our weather prophets that these are still only ideas.

11.5 Thirst

Statisticians tell us that Simon Swallower, the average man on earth, drinks annually only 5 times his weight of water, which means 350 liters (80 gallons) per year, while wheat growing in a field needs 500 times its weight in that period. However, this comparison gives a false picture, because the total water used by Simon—40 cubic meters annually—amounts to 125 times as much as he drinks, so that he uses nearly 600 times his weight of water in a year: even more than wheat does.

If Simon had to take in all his water by drinking, this would mean $4\frac{1}{2}$ liters (1 gallon) every hour, day and night, throughout the year! He certainly would burst because, even in a hot desert, the human body can lose only up to 0.6 liters (0.13 gallons) per hour.

In industrially developed countries water usage is still much higher than the world average; in Western Europe about 10 times and in the United States about 25 times the average. How could even an American Simon manage his 110 liters—about 25 gallons—of ice-cold water in every hour?

Fortunately a large proportion of Simon's usage of water does not, in fact, have to be drunk, for the total includes water for the manufacture of industrial commodities.

- for 1 kg (2.2 lb) of steel 270 liters (60 gallons)
- for 1 kg (2.2 lb) of paper 1100 liters (295 gallons)
- for 1 kg (2.2 lb) of synthetic rubber 2500 liters (550 gallons)

These figures correspond to respectively 3, 10, and 23 days' drinking for Simon Swallower.

11.6 Oil on (Troubled) Waters

If the 100,000-ton cargo of the *Torrey Canyon*, which ran aground on rocks southwest of Land's End in 1967, had been wholly spread over the sea it would have formed a layer 1 μ (0.001 mm) deep over an area one-fifth that of the North Sea; even this thin layer would have caused severe injury to fish life in the area.

An accident of this kind occurring with one of the giant oil tankers of the not distant future—say of 500,000 tons—would release enough oil to cover the whole North Sea to a depth of 1 μ; or, in Western Hemisphere terms, enough to cover one-third of the Gulf of Mexico to the same depth.

11.7 And Quiet Flows the . . . Amazon

The Amazon, the world's largest river, drains the vast and spectacular natural greenhouse, the equatorial forest of Brazil, the area of which is estimated to be three-quarters that of the United States.

Its flow of 200,000 cubic meters (7,500,000 cubic feet) per second near the ultimate mouth amounts to nearly one-fifth of all the fresh water discharged by rivers into the Earth's oceans, and the flow for 1 day would be enough to flood the whole area of the state of New Jersey (or one-half of Belgium) to a depth of 1 meter (about 3 feet).

But reverting to the cubic kilometer which is the subject of Section 1.1, even the Amazon's flow would take well over an hour to fill the space between your house and that of your friend.

11.8 The Probability of Floods

The flood which caused so much damage and loss in Holland and along the southeast coast of England in early 1953 was on a scale similar to that of the "St. Elizabeth's Day" flood more than 500 years ago. On a statistical basis it could have been expected earlier, as the calculated probability of such an event is "once in 300 years."

Now Holland's sea defenses are being so strengthened that the probability of inundation on the 1953 scale is reduced to "once in 10,000 years."

11.9 How Much of Water is Real H_2O? [57]

The association of molecules in liquid water to structures of the type $H_{2n}O_n$ is a well-established idea; in fact it seems to account for some of the unusual properties of water, steam, and ice.

The extent of this association has recently been studied, and it now appears that while earlier estimates gave about 30% for the proportion of "free" H_2O molecules in liquid water at room temperature, the proportion of this "monomeric" water may be as low as 0.1 to 0.2%.

If H_2O is the only real water, there is not much of it in the water that we see and use.

Moreover, any randomly taken sample of "pure" water is a mixture of 33 components: 18 different molecular compounds and 15 kinds of ions (water contains hydrogen, deuterium, tritium, heavy water, oxygen 16, oxygen 17, and oxygen 18).

Although these isotopes are only present to the extent of approximately one part per thousand, this nevertheless means that their total quantity on earth is 400 times what we call "the flowing water" on earth, i.e., the content of all rivers and lakes together (1.5×10^6 km^3 of isotopes); see Section 11.1.

11.10 The Tensile Strength of Water

If there were "perfect water" without any structural (temporary) flaws and without any dissolved gases, the tensile strength—calculated on basis of the surface tension—would be 5 times that of a good sort of steel.

Calculated tensile strength of water = 18,000 kg per square centimeter (270,000 lb per square inch).

11.11 Clear the Water!

Approximately 30 million people, equivalent to half the population of countries such as Great Britain, France, or Western Germany, dump their waste into and drink water from the Great Lakes at the Canadian frontier: in large areas contamination is becoming offensive.

If the addition of all pollutants to the lower lakes were to be stopped tomorrow, it would take at least 20 years for the natural flow of water through them to reduce contamination to one-tenth of the present level. In the upper districts, where the rate of flow is lower, it would take at least the lifetime of one generation for Lake Michigan to recover to that level (reached in the year 2060) and 7 or 8 generations for Lake Superior (reached in 2450).

11.12 Odd Bits

- At any time one-quarter of the earth's surface is covered with snow; 90% of all the ice is in the Antarctic.
- When rain is falling, the larger drops come down faster than the smaller because of their greater "volume (or weight): cross-section" ratio. So collisions occur, when larger drops catch up (or perhaps "down"?) with smaller ones. It is estimated that in typical heavy rain one collision occurs in every 150 meters (about 500 feet) of free fall, with an interval of about 20 seconds between successive collisions.
- The quantity of water which evaporates every year from the surface of the Earth's seas and oceans is enough to form a layer 60 cm (2 feet) deep over the Earth's whole surface.
- The ratio of the volume of sea water to the volume of air on the earth is 1:3, but the number of molecules in the sea is 250 times the number of air molecules.
- Water is approximately 36 times more viscous than air at human body temperature (37°C), but the viscosity of motor oil is 200 times as high as that of the lubricant in human joints: water.
- Powdery snow can accumulate in a very light mass with a specific gravity of 0.01; hence a layer 1 meter thick shrinks on melting to 1 cm of water (ratio 1:100).
- A man contains 70% of water.
  A shark contains 80% of water.
  A lobster contains 90% of water.
  A pipeworm contains almost 100% of water (more exactly, 99%).
- Whenever the sluice gates of the large sea-lock at Ymuiden, The Netherlands, are opened to allow a ship to pass, a quantity of salt sufficient to fill between fifty and one hundred railroad freightcars is introduced into the interconnecting waterways.

Chapter 12

Military

And yet the last war was nothing compared to what the next one is likely to be.

(Jawaharlal Nehru)

12.1 The Cost of "One Dead" in the Casualty List [58]

It has been estimated that at the time of the Romans' wars, the cost of killing one of the enemy's men was the present-day purchasing power of one U.S. dollar.

About 1800 years later, Napoleon had gunpowder at his disposal instead of the Romans' swords, axes, and catapults and the cost of one fatal casualty had risen to $3000, a tidy sum, although the average rate of increase since the time of the Romans had been a modest $1.65 per year.

In the next 100 years the cost went up fast; each man killed in World War I is estimated to have cost $20,000 (at present-day purchasing power), about 2 years' pay for a laborer in the United States. The average rate of increase (since Napoleon's time) was then about 100 times greater, $170 per year.

By the time of World War II the cost had become $50,000, thus it was more than doubled in a generation, to a sum equal to 5 years' pay for a laborer in the United States doing what we may hope is more useful work for his country.

12.2 The Cost of Arms

It is estimated that the annual costs of armaments in the United States amount to $300 per head of population, while in many "developing" countries the whole national income is still below $100 per head of population.

12.3 Wars and Battles

If we average out the declarations of wars, and the records of battles, since Columbus' discovery of the Americas, about 500 years ago, we find that there has been one official declaration of war for every 2 years of this period; one recorded battle for every trimester; and one soldier killed in action every 7 minutes.

250 wars — 2000 battles — 35 million killed

12.4 Airborne in 1870

A remarkable postal traffic was developed during the Franco-Prussian war, in 1870, between Paris, the beleaguered capital, and Tours (200 km or 120 miles distant). The outward journey was by balloon, carrying homing-pigeons, which were used for the return journey because of the limited possibilities of landing a balloon in Paris (also the prevailing wind was not favorable).

The outgoing post presented no difficulties, since the balloons could carry a payload of a few hundred kilograms, but for the return journey all text had to be printed and photographed on the smallest practicable scale, as "microfilm" to be transported by the returning pigeons. Naturally the tariff for this return post, by pigeon, was much more costly than for outgoing post.

One pigeon could transport 18 "microfilms" on which could be printed 200 pages of this book—70,000 words—for which the charge was 35,000 francs. Based on the price of butter at that time, this corresponds to 2 kg (4.5 lb) of butter per word: so one can say, that each pigeon flight was the equivalent of 140,000 kg (320,000 lb), or 10 railway wagon-loads of butter.

In all, 100 persons and 2.5 million letters were transported by these means, thus 1 letter for every 15 Frenchmen living at that time.

12.5 Explosives for the Citizen

Impressed (as are most people) by the very sound of the word "Explosives," the man in the street might be surprised to learn that the energy available from 1 kg (2.5 lb) of the well known TNT (trinitrotoluene) is not more than about 1000 kcal, comparable to the heat of combustion of a wine glass of gasoline (petrol), say 130 cm^3, or 8 cubic inches. If explosives were made to burn slowly and quietly, they might be classed as useful fuels, rather than agents of destruction.

In more familiar terms, with 1 kg of TNT, we could:

- bring 10 liters (2.2 gallons) of water to the boil
- provide the heat equivalent of the food value of 10 eggs
- supply current for a 100-watt lamp during 11 hours
- with suitably adapted engines (gunpowder was once considered as a possible aircraft fuel, in the early 1800's!) propel a small car for 1.5 km (0.9 miles) or a large American car for half that distance.

Our man in the street reaches home as a thunderstorm begins. Being well informed on weather matters, he may recall that on an average a flash of lightning releases about 100 times this quantity of energy. So, one bright flash of lightning may be equivalent to 100 kg of trinitrotoluene.

12.6 More Powerful than 150 Thunderstorms

A German author, R. Jungk, called his book about the Hiroshima bomb *Brighter than a Thousand Suns* (*Heller als tausend Sonnen*). But he might also have taken *More Powerful than 150 Thunderstorms* as its title, for the energy potential of the atomic bomb which was dropped on Hiroshima in the year 1945 (20,000 tons TNT equivalent $= 1.5 \times 10^{10}$ kcal) was equal to about 150 times the energy released in a heavy thunderstorm (see Section 6.5).

The author might also have told us that this bomb produced the same amount of energy as that released as latent heat of condensation by London's average daily rainfall (2—3 mm, or 0.1 inches) over the area devastated by the bomb at Hiroshima in 1945 (10 km^2). This is 4 times the area of Hyde Park, London, or nearly 3 times the area of Central Park, New York. The effect of the explosion was so enormously greater, because its energy was released in but 10^{-8} second, compared to a whole day for the rain.

Indeed, the heat of condensation annually released by the formation of raindrops the world over totals the same energy as would be released if every one of our fellow inhabitants were to drop 6 "Hiroshima" bombs.

The most powerful rocket of World War II was the German V-2, which was propelled by the combustion products of alcohol. Its explosive warhead, 820 kg (1800 lb) of TNT, was sufficient to produce, on impact with the ground, a crater of which the volume was about that of a dwelling house—3 meters (10 feet) deep, and about 13 meters (45 feet) across.

This energy is equaled by the heat of condensation of 3 or 4 day's average London rain on the area of the crater, about 130 square meters or 155 square yards, that is, the area of the roof of a house.

12.7 A "Blockbuster" for Everyone [74]

At the end of World War II, German cities were being attacked with the heaviest "ordinary" bombs then developed, which were capable of destroying rows upon rows of houses and other buildings at one blow. With their rated power of 15 tons TNT they well earned their name "blockbusters."

But the power of the atomic bomb which was dropped in that same year, 1945, on Hiroshima was rated at 20,000 tons TNT, which, in terms of "blockbusters" would have been enough to equip every inhabitant of a small village—1300 people—with one blockbuster each.

Hydrogen bombs have since been developed with powers up to 20 million tons TNT, which corresponds to one blockbuster for each of the 1 million inhabitants of Brussels, and in 1961 the Russians are believed to have detonated a bomb in Nova Zembla with a power sufficient to provide every resident of Boston (4 million population), with the means of destroying a large part of a small town at one blow.

By 1962, the United States atomic stockpile was enough to equip Asia and Africa—2000 million people—with a blockbuster each, and it is now estimated that the stockpiles of the United States and Russia, combined, rate a total of 60,000 million tons TNT, more than enough to supply one blockbuster for each one of the Earth's inhabitants.

A Propos

To equal the explosive force of an atomic bomb containing a charge (fissile material) the size of a tennis ball, a quantity of dynamite would be needed sufficient to pack Times Square (New York) or Piccadilly Circus (London) to the height of the surrounding buildings.

12.8 Highlighting the Enemy

In the endeavor to spot guerilla fighters on the move at night in

Vietnam, the Americans have fitted lights of 6 million candlepower under some of their military aircraft. With these, they can—from a flying height of 3300 meters (10,000 feet)—illuminate an area of 800 km² (300 square miles) with 4 times the brightness of the full moon. This is about the area of Greater London, or about half the area of New York and its "satellites."

From the lower flying height of 600 meters (1800 feet) they illuminate about 6 km² ($2\frac{1}{4}$ square miles)—an area 3 times that of Regent's Park, London, or about that of Manhattan from The Battery up to 33rd Street (the Pennsylvania R.R. station)—at 60 times the brightness of the full moon.

But this is negligible, compared with the project Able under development in the United States, which aims at reflecting sunlight from inflatable mirrors, 700 meters (2000 feet) in diameter. It could light up an area of 100,000 km² (40,000 square miles), larger than the state of Indiana or the country of Greece with a brightness equal to twice the moonlight.

12.9 Pain-Killing Medicine

Authorities in the United States have estimated that in the event of nuclear war, 65 million people would perish, a number about equal to the present population of West Germany.

For the relief of pain caused by "nuclear" injuries it is believed that the equivalent of 1 kg of opium would be needed per victim. Assuming that the same number of nonfatal victims shall be added, this would mean the cargo of 10 medium-sized freighters carrying only opium (1 freighter = 13,000 tons).

12.10 The Daily Burden of Soldiers [59]

The various methods of carrying a bag vary little the world over; nevertheless the alternatives seem to be related to social standing.

Tests including pulse records, breathing rates, and analysis of expired air were carried out on soldiers as they walked carrying a bag in various ways, showing great differences in efficiency.

- The "rucksack" method, as used by schoolboys, hikers, and of course the Army, turned out to be by far most effective.

- The "haversack" method—the bag hung from one shoulder—involved 18% more energy.
- The "attaché-case" method—the bag held in one hand—90% more energy.

So, as in many other cases, military men and schoolboys—not to forget hikers—have developed a more effective method of carrying their daily burden than that customarily used by our diplomats.

12.11 Odd Bits

● One result of the world's outlay on military strength—about $150 million, approximately £62.5 million, or $40 (£17) per head of population, annually—is that the world's armies comprise no less than 25 million men, a number equal to the whole population of a country such as Canada.

● Military power is probably at its strongest in a very small, though important, state, the Vatican. Of the officially recorded population of 500 odd, 40% are military personnel, including the famous Swiss Guard, and "only" 35% are priests and higher clerics.

● It is said that in World War II, only 40% of the G.I.'s in combat fired their weapons; in Korea 60%, and in Vietnam 80%.

● As long ago as 1937, the victims of traffic accidents in the United States in that one year numbered twice as many as the American soldiers killed in the First World War.

● During the last World War the Americans built 50 synthetic rubber plants: the capital involved was equivalent to 3 days of warfare (750 million dollars).

Chapter 13

Sport and Games

It is, indeed, a pleasure to win from You, my dear fellow, and to say this is the highest praise I can bestow on a man.
(Tiberius Claudius)

13.1 Sport and Life on Various Levels

One of the lowest levels of activity for an average man is lying flat on his back, i.e., "supine," perhaps "under the spreading chestnut

Lying flat on his back (1700 kcal per day and night).

tree." This will call for an energy consumption of 70 kcal per hour, which adds up to 1700 kcal for a day and a night, some 30% below the 2500 kcal required every day by a man living a normal life.

This is however not the very lowest level possible, because in the "frozen" condition, as now used in advanced medical or surgical procedures, a man's daily energy consumption is considerably less than the normal 1700 kcal; perhaps deep-freezers offer hitherto unforeseen possibilities to a modern "lazybones."

More interesting than such idle persons is that well-known all-rounder sportsman, Mach A'Mover, who was also able to obtain some accurate measurements of his energy consumption corresponding to various quotas of daily activity:

- Pacing to and fro in the gymnasium at 5 km (about 3 miles) per hour, Mach found that his energy consumption during this period of exercise was $4\frac{1}{2}$ times his needs when in the supine position. So, a good 4 hours of walking in this manner, followed by 20 hours of lying flat, would mean a consumption equal to the human daily average, 2500 kcal per day.
- On an outdoor training walk at $6\frac{1}{2}$ km (4 miles) per hour the energy consumption rose to 7 times the "supine" needs, but Mach felt that he would like to go faster, so took his bicycle. However at just over twice the speed, 15 km ($9\frac{1}{2}$ miles) per hour, the energy needed was less than for the training walk, being only 6 times the needs when supine.
- Now winter comes, and, substituting a pair of skis for his pair of wheels, Mach finds even a short excursion, gliding over the snow, very tiring. He might well, seeing that the energy consumption was up to $8\frac{1}{2}$ times the "supine" needs; and, as walking on hard snow at only 6 km ($3\frac{3}{4}$ miles) per hour means an even higher consumption of energy, 10 times the "supine" needs, he decides that skiing is about at his limit of activity.
- His brother, Mick A'Mover, was of a different caliber. Hearing that rowing—at "33" (strokes per minute)—would result in energy consumption at 15 times the needs for the supine position, he decided that this was the ideal sport, at least for the summer. In winter he could take his exercise by walking in soft snow, using snowshoes; even at 4 km ($2\frac{1}{2}$ miles) per hour this would mean 12 times the supine energy needs.

Moreover when the weather was too bad for any outdoor activity, he could still utilize the energy from his hearty meals by boxing—at 7 times the supine energy level—or dancing at, at least, 4 times the supine energy, depending further on the style of the dance.

- The brothers had forgotten, so far, that even their basic supine position required certain incidental activities for which the energy needs were:
    - Standing at (not leaning against) the bar, or light work at the office 1.7 times supine needs

- Washing the hands, or driving the car 2.3 times supine needs
- Washing the dishes 2.8 times supine needs
- Walking upstairs at 1 stair per second 4 times supine needs

Neither had either of the brothers A'Mover, Mach and Mick, considered that making their beds was just about 6 times as tiring a task as lying on them.

13.2 Weight Lifting: 1

The greatest weight so far lifted by a human being is 2720 kg (6000 lb), which was achieved in the United States at Toccoa, Georgia, a location on about the same latitude as Gibraltar. This was a weight exceeding that of the heaviest Rolls-Royce car, 2540 kg (5600 lb).

In the same competition, another weight lifter succeeded in raising just such a Rolls-Royce car, together with chauffeur and "groom," but the record remained unbroken as the weight of these two men was 180 kg (400 lb) so only enough to bring the total up to the "record" figures, 2720 kg or 6000 lb.

But, after a good dinner with a physicist, the aspiring challenger persuaded the umpires to let him repeat his attempt on the record at the equator, using a brand-new spring scale for the weighing, and incidentally promising the umpires a nice trip.

In order to avoid any misunderstanding, the chauffeur and groom again took their places in the Rolls-Royce, and feeling sure of his success, the challenger added a spare tire, making an extra 5 kg, or 11 lb. All went well and it seemed that a new world record of 2725 kg or 6011 lbs had been established.

However, in spite of their enjoyable trip to tropical waters, the umpires "were unconvinced and decided to have the performance repeated at the North Pole using the same spring scale. To his chagrin, the challenger could not lift the car and its occupants, even after the spare tire had been removed; indeed two more tires had to be taken off the wheels before the poor fellow could achieve his lift, of a weight now 2709 kg or 5976 lb!

Becoming suspicious about the laws of gravity and the action of the spring balance, the President of the Board of Umpires came to the conclusion that the car should be weighed at all three places, Toccoa, the equator, and the North Pole using legally attested weights on an ordinary weighbridge; the results were that the apparent weights at all three places were the same.

The President, following the advice of one of the umpires, also decided to invite a physicist to dinner, and so became aware that "mass" and "weight" are not the same thing. He learned, in fact, that while the mass of an object is the same anywhere in the universe, its weight, being the force by which an object is attracted to another—usually larger—object, such as the Earth, depends on the mass of this second object, and on the distance between the two objects.

With this second point in mind, and realizing that one is further from the Earth's center when standing at the equator, than when one is shivering at the North Pole, it became clear to him why the effective weight of the Rolls-Royce and its occupants was less at the equator than at the North Pole. "Moreover" the physicist had told him "there is an appreciable lessening of effective weight at the equator by reason of the centrifugal force caused by the Earth's rotation, which indeed tends to throw objects whirling round the Earth's axis—though on its surface—into the depths of space."

Place and latitude		Acceleration owing to gravity (in meters per sec^2)
Toccoa (Georgia) and Gibraltar	35°	9.80
Equator	0	9.78
North Pole	90°	9.83

13.3 Weight Lifting: II

It is also on record that in Tampa, Florida, a woman lifted one end of a $1\frac{1}{2}$-ton car, under which her son was trapped, in a desperate effort to save his life.

H. G. Wells might have used these facts in his story "The First Men in the Moon"—one imagines Cavor, one of the travelers in the "anti-gravity" capsule which Wells described in such convincing detail, astonishing the Selenites with accounts of how men from his own planet would be able, on the moon, to lift a vehicle for 4 passengers and some luggage with one hand and hold it above his head.

For, on the moon, the car would weigh—for the man from Earth—only 60 kg (130 lb).

But, extending Cavor's travels to Jupiter, he would have cut a much less striking figure; indeed if he had taken the champion weightlifter with him (and his "Earth" weight), the best to be expected on Jupiter would be a lift of the weight labeled 140 kg (300 lb), less than half what has been achieved on the Earth.

13.4 Running Records

In the last hundred years of athletic competitions, the "time" records in the mens' races have fallen, steadily or otherwise, for all lengths, and it is estimated that by 2000 A.D. these records will have fallen still further, at about 0.2% per year.

If so, the record for the mens' mile would then be 3 minutes 41 seconds, corresponding to 26 km (16.5 miles) per hour. A winner with that time to his credit might be disappointed to learn that this speed is no more than a fair average for the African elephant, not one's first thought in considering speedy animals.

It is more encouraging to discover that the sprinting record for men (100 yards or meters), equivalent to 36 km (22.5 miles) per hour, means a speed almost equal to that of a running deer. If the same annual improvement of 0.2% holds for short distances, then by the year 2000 A.D. our best sprinters should be just about equaling the deer's performance at 38 km (24 miles) per hour.

13.5 What a Man Can Do

A man (world record) can:

- swim (100 meters), 1 time his own length per second
- run (100 meters), 5.6 times his own length per second
- high jump 1.2 times his own length
- long jump 4.5 times his own length
- putting the shot 10.8 times his own length
- discus throw 33 times his own length
- hammer throw 39 times his own length
- javelin throw 48 times his own length

But a flea, restricted by Nature in its range of possibilities, excels in two respects:

- high jump 50 times its own length

- long jump 80 times its own length

Its long jump reaches relatively twice as far as our hammer throw.

13.6 Anaerobic Sportsmen or Sportswomen [60]

For an individual performing physical work, the subjective discomfort is related to the oxygen usage which the work entails rather than to the work output as such. Based on this, arbitrary scales have been arrived at giving a measure of "work sensation."

However, such scales fail completely in the case of a 100-yard sprinter who may take literally no breath at all during the 10 seconds' duration of the race; that is to say, he—or she—performs anaerobically. As a result of this the runner incurs an "oxygen debt" which is paid back by increasing the breathing for several minutes after the end of the race.

During the anaerobic period he—or she—derives energy from the conversion of glycogen stored in the muscle cells. The maximum output available from this source is some 28 kcal, which, in terms of another realm of sport, should be enough for a weight lifter to raise 25 kg through 1 meter 40 times in 4 minutes.

In practice the efficiency of the processes whereby the "oxygen debt" is repaid does not exceed 25%, so that using only the energy stored in the muscles, the weight lifter could do his lifting in only 1 minute, *if* he were able to carry on for so long anaerobically.

13.7 Breaking One's Neck

When a large 1500-kg (3300-lb) car traveling at the speed of 100 km (65 miles) per hour is braked to a standstill, the mechanical energy absorbed by the brakes is just enough to evaporate a half-pint of water (250 cubic centimeters).

An Olympic sprinter, suffering the misfortune of colliding against a wall with "his" velocity (100 meters in 10 seconds, or 36 km per hour), loses not only his life but also an amount of energy sufficient to evaporate a mouthful of water: 15 cubic centimeters.

13.8 Climbing on the Eiffel Tower

A motor manufacturer's folder suggests that the energy needed to climb the 300-meters (1000-feet) high Eiffel Tower on foot would be about the same as the energy expended in changing gears during 3 hours.

But we do not recommend climbing the Eiffel because police reports show that in the 75 years of the Tower's existence (during which period a population equal to that of France have ascended it) 350 persons have jumped from it to their deaths. If the whole population of the world had climbed the Tower and the same proportion of visitors had taken this fatal step, the horrifying plunge would have been just a daily spectacle: one victim a day!

13.9 Cycling to Mars

When the Americans' "Mariner IV" sent back close-up photographs of Mars, in 1965, it occurred to a keen cyclist to reckon when he would have had to set out on his machine to cover a distance equal to that traversed by Mariner IV on its way to Mars. He found that, even riding night and day continuously, each and every day, he would have to have started at the time of the building of the Pyramids in Egypt, that is about 3000 B.C.

13.10 New Standards for Championships Adjusted to Space Traveling

When we will have established ourselves on the Moon or on other planets, sports records will need a new look. In fact, what we now call a "world record" will have to be more exactly described as a local "*Earth* record."

Disregarding those differences arising from the atmospheres themselves, that is atmospheric pressure, humidity, oxygen content and the like, let us imagine competitions taking place in large halls within which the Earth's atmospheric conditions are maintained, so that we are dealing only with differences between the gravitational forces.

On the moon, of which the diameter is one-quarter that of our planet, the force of gravity is only one-sixth that by which we are attracted. As a result—contradictory to what seems obvious—walking pace becomes slower because the body, which in walking is repeatedly raised about 4 cm, will drop back more slowly on the moon than it does on the Earth. So we shall be kept to a snail's pace while strolling around on the moon. Someone has calculated that in a walking race on the moon the maximum speed would not be more than 10 km (6 miles) per hour, which is about the speed for the Earth record for swimming.

On the other hand the achievement of a high-jumper will, in the nature of things, be enhanced. So we will be able to jump up as high as a kangaroo and then return to the surface quite slowly. The Earth record of 2.15 meters (7 feet $5\frac{3}{4}$ inches) height would be well beaten with a jump of about 14 meters, that is nearly three times the Earth record for the pole vault!

13.11 Sport and Nicotine [61]

"Thanks," said James to the offer of a cigarette, "I will stick to my pipe. But my brother John will be glad of one: he smokes a packet a day. And, I might add, he heard an interesting tale the other day from his doctor about the effect of smoking only half a cigarette."

After smoking this "half" the doctor said: the effects on the pulse rate and on blood pressure can be detected for about 25 minutes, though they are at their maximum about 4 minutes after "lighting up." In numerical terms these maximum effects are:

- increase in pulse rate from say 65 to 85 per minute;
- rise in blood pressure from say 108 to 115 mm mercury (Hg).

  If such a rise of 6.5% should occur in atmospheric pressure, this would mean a change from storm (730 mm Hg) to nice stable weather (775 mm Hg).

His doctor termed the product of these two, pulse rate and blood pressure the "heart-loading" index—it parallels the body's consumption of oxygen.

Averaged out over the whole 25 minutes during which the effects can be detected, the rise in index is about 20%—from 7000 to 9500 units—and this is the result of an intake of only 3 mg of nicotine which is present to the extent of about 1.3% of the weight in most cigarette tobaccos. Interestingly enough, the effects are less with nonsmokers; for them—if they can be persuaded to enjoy a "whiff"—the rise in heart-loading index from smoking half a cigarette would be about 10%.

The doctor illustrated his point with some simple examples of physical effort which would impose about the same extra loading on the heart and coronary system as smoking the half-cigarette:

- cycling for 20 minutes against a light head wind
- supplying the power needed for a 20-watt electric lamp, a light hardly bright enough to read by
- finding the lift out of order on one's return home, and climbing, albeit at a moderate tempo, the 120 stairs to a flat in one of the new "tower" blocks.

Fortunately, smoking a cigarette down to quite a small butt increases the heart's load by only 25% above that resulting from smoking the first half of the cigarette. Even so, the daily packet means an extra load on the heart about equal to the effect of cycling for 8 hours against a light head wind.

The doctor concluded with a comparison of cigarettes, pipes, and cigars, per gram of tobacco; the effects are about the same for pipe smokers as for cigarette smokers, but for a cigar, the effects are about 25% less.

13.12 The Golfer's Diet

Human metabolism slows down with advancing age; consequently, if the intake of food remains about the same, the older man needs more exercise than the younger to maintain a slim figure, and in terms of golf this means that a young man should plan his playing somewhat as follows:

at age about
- 20 two rounds of 18 holes, and 6 more holes for practice
- 30 two rounds of 18 holes, and another half-round (9 holes) for practice
- 40 three rounds; however, ending the last at the fifteenth hole
- 60 three rounds, all the way

13.13 The Inside and the Outside of a Golf Ball

A good shot can easily result in a traveling away of the ball from the club head with the velocity of a fast train: 150 km (95 miles) per hour, but long hitters may produce speeds of 250 km (160 miles) per hour.

The normal depth of the "dimples" in the cover is 0.3 mm (0.013 inch). It is estimated that if they were only one-tenth of this depth, the flight to "first bounce," other things being equal, would be halved.

The dimpled surface of the ball also helps to get "back-spin," which is said to amount to some 60 revolutions per second from a good tee shot. This is not far behind the revolution speed of the motor in a small automobile (6000 rev per minute = 100 per second). But a 9-iron shot at 170 revolutions per second is no exception. Such figures may look surprising to a golfer, but perhaps the confidence in his own ability is restored when he calculates that in his 200-meter (220 yards) drive the ball will travel 1 meter (3.3 feet) before it has completed one revolution.

In your speech at the golf club's annual dinner, ask how many dimples there are on the cover of a golf ball. The conservative diners may say 100, and the progressives, perhaps, 200, but there are actually 332.

Secondly, what is hidden inside?

Inside of the golf ball is a thin rubber thread, stretched to the length of a good tee shot, 255 meters (285 yards); compare this with the world's record drive of 390 meters (430 yards) or with a somewhat poorer tee shot with a carry like the overall length of the Queen Elizabeth or the France, world's largest liners (about 320 meters = 345 yards).

This stretched thread of rubber exerts a force on the inner core of the golf ball which exceeds the weight of a large car, being about 3000 kg (7500 lb): the energy stored in the thread is enough to raise a man weighing 70 kg (150 lb) off the ground to a height of 60 cm (2 feet).

13.14 Peter Player's Chances

● *Bridge*

With a smile Peter Player laid his contract hand, card for card, on the table . . . all thirteen spades! "I take it you'll concede my slam, gentlemen?"

Dr. Statico, one of his opponents, wouldn't have been a statistician, if he had neglected an opportunity to discourse on his pet subject—probability:

"My dear Peter, I can offer you only little hope of an immediate repetition of your abominable luck. Even if you played forty hands of bridge every evening, you would need—mathematically speaking—to have commenced playing with the primitive types of prehistoric mammal (11 million years ago), since, at that time, *Homo sapiens* had yet to arrive on the scene. Then you would have had a statistically justified expectation of personally drawing all thirteen cards of *any* suit."

"But if you are going to insist on *spades*, Player, better choose another example to bring home to you the whole somber truth. If every living person in the world (3500 million) participated in a marathon drive of two hundred hands, only one of this giant company of odd players could be reasonably expected to draw a complete suit of cards. Not until the marathon had been played four times over would there have been a statistical likelihood of any complete suit drawn being *spades* rather than one of the other three suits. And then . . . if it is precisely you, dear Peter, who is to have a reasonable expectation of finding all the spades in your hand, you must be prepared to attend over 3500 million such marathons!"

"But I can tell you more about the statistics of this marvelous game and even in precise terms. The number of possible bridge hands is exactly

635 013 559 600

that is at least 6 times the number of celestial bodies in our Galaxy" (see Section 3.14).

● *Chess*

Realizing that such astronomic figures did not appeal too much to his bridge friends, Dr. Statico refrained from telling them that the number of possible games of chess is estimated to be $10^{10^{50}}$, a number beyond computation and perhaps nothing compared with the number of cubic millimeters in the observable universe.

● *Roulette*

Therefore Dr. Statico confined himself to a much simpler story. Knowing that the longest consecutive run of even numbers in the history of Monte Carlo is a series of 28, he told the bridge group that a roulette wheel should have started to spin "no more than" some 3000 years before the Pharaohs reigned (altogether 6500 years ago) to give a statistically justified expectation of a sequence of 29 even numbers occurring once.

13.15 Communism through Gambling

In 1962 the British public gambled on the outcome of sporting and other events to the extent of £15 ($40) per head of population, which worked out to well over £850 million ($2335 million) in total. This prodigious total of stake money, 65% of which was wagered on horse racing and 10% on football, is about twice the amount annually devoted by the United States to foreign-aid programs.

No wonder that a famous man once called his poker club: "The Society for Equalization and Distribution of Wealth."

13.16 Odd Bits

- The leaping nematode, an unsegmented little worm, living in Australia, bridges gaps by leaping about 10 mm, which for a human athlete would be about 10 meters (30 feet), surpassing man's world record in long jumping of 8.4 meters (25 feet).
- The average busy housewife spends about 15 hours every day walking about, and in this time covers a distance of nearly 25 km (16 miles).

Chapter 14

Surprises for the Great

What is new today may be obsolete tomorrow.

Uranus, mother Earth's "husband," would be surprised to learn the following concerning his wife's future. She did store so much of his stuff—uranium—that, although 90% of the original U-235 has been converted into U-238 due to radioactive decay at a rate of 1.2×10^{10} atoms per second, mankind will presumably be able to cover the gap in its energy supply (coal and oil will be exhausted in a few generations) until new energy sources can be developed (see Section 6.1C).

Zeus, the chief of the Greek gods in classical times, would be surprised to learn that if all the water in the earth's oceans were to be spread over what is now land, the gods in their abode on Mount Olympus—3000 meters (about 10,000 feet) high—would see the waters rising almost to their feet.

Plato (400 B.C.) would be surprised to learn that at every breath we are likely to inhale some 50 million molecules of air exhaled at any time by this great philosopher—or by any other person who lived for 65 years—and whose breath has been well distributed throughout the atmosphere (as would be expected to be the case after so many centuries).

Aristotle (350 B.C.) would be surprised to learn that the moon, according to his theories a perfect sphere (as were all nearby heavenly bodies), is in fact mountainous in many parts. Modern astronomy, and especially the information sent back by spacecraft on or near the moon's surface, indicate that the moon's mountains may well be up to twice the 3000-meters height of Mount Olympus.

Alexander the Great would be surprised to learn that his death in 323 B.C. was caused by the bite of a malaria mosquito infected with a parasite which can multiply so rapidly that within 24 hours each cubic centimeter of his blood might have contained 300 million offspring.

Eratosthenes, who calculated the circumference of the earth in 200 B.C. by measuring the difference in altitude of the sun at the same moment at two different places (Alexandria and Assuan), would be surprised to learn that the error in his calculation was no more than the distance between New York and Philadelphia (200 km), only 0.5 percent of the actual circumference.

Aristarchos of Samos, living in the third century B.C., was the first to put forward the idea that all the planets, including the Earth, move around the sun.

He would now be surprised to learn that the movement of one of his fellow islanders through space includes at least three separate rotations and a fourth "drift":

- rotation about the Earth's axis, which for the latitude of Greece is just a little more than the speed of sound ($\frac{1}{3}$ km per second, or 1100 feet per second);
- rotation of the Earth about the sun, which means movement at about 100 times the speed of sound along the Earth's orbit;
- rotation of the sun and solar system as a whole about the center of our "Milky Way" Galaxy, movement at about 600 times the speed of sound;
- finally, a drift in the direction of the constellation "Hercules" at about 60 times the speed of sound.

Ptolemy, who in the second century A.D. estimated the distance from the Earth to the sun to be 7 million km (4.5 million miles) would have been surprised to learn that during the reign of Louis XIV of

France, around 1700, astronomers came to the conclusion that the distance to the Sun was about 20 times as much.

If Ptolemy remembered his people's invention of "democracy" he might have been still more surprised to learn that Louis in his style of "Roi Soleil" was happy to hear from his learned advisers that if indeed he should take up residence on the sun, instead of at Versailles, he would hear a cannon shot, fired at the time of his birth, only after the lapse of 20 years—a demonstration of the wide gap separating him from those he reigned over, who must all have heard the shot within half an hour (supposing that it was loud enough).

Trajan's reign was in the early years of the second century A.D. He would have been surprised to learn that the water supply of Rome at that time was already one-third of the supply available to the city in 1950.

Mohammed, who founded the Islamic faith about 600 A.D., would have been surprised to learn that the number of believers joining in the ceremonies of the seventh day of the "Hadj" pilgrimage, held near Mecca, is now only 70,000. In Mohammed's time there was an assembly of 700,000, and tradition tells us that it was then foretold that any falling off from this number would be made good by the attendance of angels.

Johann Gutenberg, who is considered by many historians as the inventor of bookprinting (about 1400), would be surprised to know, that there exist now some 5000 daily newspapers, the largest of which is Ahasi Simburu in Japan, distributed in as many copies as there are inhabitants in New York City (over 10 million).

Christopher Columbus, the great fifteenth-century navigator, would have been surprised to learn that about 450 years after his voyage to the "New World," the government of an important part of that world—the United States—would be approving a budget of $40,000 million for the project of landing on the moon (this amount not including the preliminary research expenses) which involves a journey no more than 65 times as long as his own pioneer voyage to the New World.

He would feel that his journey, costing $6.00 per mile for the 3 ships carrying 120 people, was much better value than the projected trip to and from the moon, costing $17,000 per mile (return) for only a few fellow explorers for whom a successful landing was a matter of some uncertainty, not to mention the difficulties of effecting a return trip.

Johannes Kepler (1571–1630), who developed the classic laws concerning the rotation of planets about the sun, would be surprised to learn that now approximately 1000 artificial satellites have been launched from our globe, part of which are still in use for communication purposes.

Around the year 1600, *Galileo* succeeded in making a telescope with magnifying power about "4×," and with this he could detect stars which we now describe as "of the 6th magnitude"—that is, not very bright. He would have been surprised to learn that 350 years later men would make a telescope—the one on Mount Palomar, in California—of which the mirror is 5 meters in diameter. With this giant, optical records can be obtained of stars of the "20th magnitude," which have only 1/400,000 the brightness of the 6th magnitude which he could just see using his seventeenth-century instrument.

Otto von Geuricke of Magdeburg, Prussia (mid-seventeenth century) would have been surprised to learn that when his experiment with the two evacuated hemispheres was studied by his scientist successors some 300 years later, it was realized that the efforts of the 8 strong horses were opposed by two armies of rapidly vibrating molecules, beating on the hemispheres from the outside, and each numbering—at any given moment—1000 times the population of the Earth in the 1960's, amounting to about 3500 million.

Christian Huyghens was one of several scientists who, around 1650, studied the speed of sound and arrived at a figure reasonably near 340 meters per second which is now accepted. He would have been surprised to learn that 300 years after his experiments men would be able to see large objects traveling through the air at this speed—in the form of a supersonic jet aircraft, as it flashes overhead.

James Watt reported that a bushel—about 38 kg (85 lb)—of best

Swansea coal would yield, when used in the boilers and engines of the late-eighteenth century, about the same amount of work as 10 horses could do in an hour. He would have been surprised to learn that if the coal could have been burned with 100% recovery of the heat of combustion, and the engine operated at 100% efficiency, 50 times as much work could have been obtained from the bushel of "best Swansea coal."

George Washington would be surprised to learn that the population number of the United States today is equal to that of Europe when he died in 1799 (190 million).

Napoleon would be surprised to learn that he might have read in the *London News* of April 27, 1968:

> ". . . there was never any compelling reason to abandon the duodecimal for the decimal system. The former is in many respects more accurate. A third of a present pound is 80 pence or 6 sh. 8 d., but a third of a decimalized pound will be £33.333333333 . . . 3 *ad infinitum* 'new' pence. We will be fully decimalized in 1971 just over a century and a half since the decimal system was adopted by revolutionary France on the grounds that the new is necessarily better than the old."

The astronomer *Sir William Herschel* who, around 1800, made the first serious attempt to estimate the total number of stars in the universe (though his method in fact only covered our "Galaxy") would have been surprised to learn:

- that the Galaxy could contain 100 million times more of our suns than the number of drops of water in all the Earth's seas and oceans;
- and that we must multiply this 100 million again by 1 million to express the number of times our Galaxy could contain the volume of the Earth.

E. I. Du Pont de Nemours, who bought in 1802 a little piece of ground to start an industry with a capital of 36,000 dollars and employed at that moment 17 people, would be surprised to learn that "his" company now has annual sales equivalent to about one dollar for every person on the Earth.

John Dalton, who, around 1800 propounded his "atomic" theory so important in the later development of chemistry, had to reduce the admission charge from 10 shillings to 5 shillings for a course of 12 of his lectures—and even this failed to improve the attendance. He would have been surprised to learn that 150 years later a high academic figure, for example a professor, would expect to receive at least 30 guineas (about $100.00) for only one lecture.

George Stephenson whose steam locomotive, the "Rocket" inaugurated in 1830 the age of railways by traveling between Manchester and Liverpool at 55 km (35 miles) per hour would be surprised to learn that:

- recently in Japan a railway as long as the distance between Manchester and Aberdeen, or between Washington and Boston (515 km) was built in 6 years time for trains with a speed of 250 km per hour, whereas the French achieved in experiments speeds of 330 km per hour, 6 times that of Stephenson's locomotive;
- present rockets move more than 400 times faster (25,000 km per hour) than Stephenson's "Rocket" did.

Frédéric F. Chopin (1810–1849) would be surprised to learn that now the record for playing the "Minutewalz" is striking the piano keys at a modest 20 times a second—the rate at which locusts flap their wings.

Justus von Liebig, a chemical pioneer of the mid-1800's, might have been surprised and would surely have been pleased to learn that the school of chemistry which he founded counts 34 Nobel Prize winners among its alumni.

Ernst Mach (1838–1916) who presented in 1887 his famous paper on shock waves, would be surprised to learn that aviation developed so slowly, that it took 40 years before the Mach Number (320 meters per second = 1200 km per hour) was generally recognised as an international standard and that another 30 years lapsed before (military) aircraft broke the sound barrier.

F. M. A. Dumont who took out a patent in 1851 for "a particular combination of electric wires for the conveyance of intelligence in the interior of large towns"—this was 25 years before the telephone was invented—would be surprised to learn that the number of telephone subscribers in the world now surpasses the number of American citizens (over 200 million subscribers).

Ernest Solvay, who in 1863 discovered, with his brother, the process for transforming ordinary salt into washing soda now known as the "ammonia soda" process, would have been surprised to learn that about 100 years later the world use of soda would be at the rate of 5 kg (11 lb) per year per head of population.

The Russian chemist *Mendeleev* (or Mendelejeff), whose conception of the "Periodic Table" of the elements in about 1869 remains valid, though much extended, 100 years later, would have been surprised to learn that his idea would lead to the discovery of not only all the elements up to and including uranium, with atomic number 92, but also of elements of still higher atomic number and of isotopes, some of familiar elements and many of extreme rarity or short life—in total about 1200 "novelties."

The painter *Vincent Van Gogh*—about 1870—sold his picture "Man with the Pipe" for no more than the price of a packet of tobacco. He would have been surprised that less than 100 years later the picture would command a price high enough to supply a pipe smoker with his tobacco for a thousand or even ten thousand years, for an economical smoker, if the taxes on tobacco remain as they now are.

John D. Rockefeller, who founded and built up the Standard Oil Company at the end of the nineteenth century, would have been surprised to learn how far the atoms in his petroleum products might travel.

The gases from a car (principally water vapor and carbon dioxide) produced over only 1 meter of travel (about 1.1 yards) form a cloud which rapidly disperses and is diluted in the Earth's atmosphere. But if John D. found himself as far from the Earth as the moon's distance, he could still expect to find—outside his spacesuit—a few molecules* of

*Provided, of course, that no condensation to liquid water or other loss of material occurred.

the exhaust gas from every motor car which had been in use on the Earth's surface some little time before.

On the other hand he would have to realize that the whole car population of the Earth, about 130 million, would have to be driven 1 million times round the Earth before he could expect to see a drop of water formed from the combined exhausts of all these cars, even though the water content of exhaust gases is about two-thirds of the whole.

Louis and Auguste Lumière, who, in 1895, in a small cellar in Paris, showed the first animated film, would be surprised to learn that in most developed countries—including Russia—there are now about 5 cinema seats per 100 inhabitants.

The brothers *Orville and Wilbur Wright* who made in 1903 the first powered flight at Kitty Hawk (North Carolina) of 59 seconds would be surprised to learn that now at any moment thousands of people fly from one distant town to the other at 1000 km (650 miles) per hour.

Konrad Zuse a pioneer in the construction and operation of computers of the modern type, who had a machine operating in Berlin in the year 1941, would have been surprised to learn that the 1500 valves of his equipment have now been replaced by integrated transistor circuits and that some 20 years after his work, computers are recognized as one of the main factors leading to the "Second Industrial Revolution," now well under way.

Enrico Fermi, who set up the first simple atomic reactor in 1942, would have been surprised to learn that 25 years later, "breeding"—that is "self-supporting"—reactors are being developed which will produce electricity at a rate in megawatts equal to their annual requirement of fissile material in kilograms. So, 400 kg fissile material will soon be sufficient for a 400-MW plant, which in turn would be sufficient to supply 1 million of the inhabitants of a highly developed country.

In 1887 *Michelson and Morley*, eminent physicists, were investigating the velocity of light to an extreme degree of accuracy; in fact they were looking for a difference in its velocities along, and across, the direction of the Earth's motion around the sun. For reasons now well understood, they found no difference, but they would have been surprised to learn that some 80 years later, aircraft would be covering quite appreciable distances, upwards of 40 meters, in that small fraction of a second which light would need to circle the Earth.

Guglielmo Marconi, who filed the first application in Britain for a "wireless" telegraphy patent in 1896 would have been surprised to learn that 70 years later physicists would have developed the "Laser," from which a single beam can, in theory, transmit information at a greater rate than all the "wireless," that is "radio" channels now in use.

John Boyd Dunlop, whose factory started to make pneumatic tires also in 1896, would have been surprised to learn that a tire looses a layer only one atom thick, on average, during one revolution of a wheel, so not more than 500 atoms thick in each kilometer (or 800 thick after running a mile).

Lord Rutherford, who in the early 1900's discovered the spontaneous transmutation of elements based on radioactivity, would be surprised to learn that now, after some 100 years of intensive industrial development, man has altered—albeit very slightly—the natural balance of radioactive carbon, ^{14}C, to ordinary carbon, ^{12}C, in the atmosphere by burning immense quantities of fossil fuel—coal and oil—formed many million years ago. This fossil fuel now contains less ^{14}C than the proportion in the atmosphere when it was formed so that there is less ^{14}C in the atmosphere than hitherto. It is even thought that freshly grown wood, if used to imitate old Jacobean style furniture of 200 or 300 years ago, could, for this reason, "pass" the carbon-dating test by matching the present ^{14}C proportion in wood from a tree felled in 1700.

Bibliography

GENERAL SOURCES

Encyclopaedia Britannica

Encyclopedia Americana

Winkler Prins Encyclopaedie, Elsevier

The Guinness Book of Records, London, 1965

Basic Statistics 1966, European Communities, Brussels

[1] The illustrations on pages 56 and 176 were borrowed from BRES, The Hague.

[2] Contributed by J. P. van Lier, M. Vertregt, and R. E. J. Zieck.

[3] *The Giant Colour Book of Mathematics,* Paul Hamlyn, London.

[4] J. H. Fremlin (private communication).

[5] M. Rothschild, *Scientific American,* Dec. 1965, p. 44.

[6] Neutrino Conference, CERN, Geneva, 1965.

[7] J. van Diggelen, *Natuur en Techniek,* Feb./March, 1966.

[8] *Science Progress,* June 1964, p. 202.

[9] R. P. Fenkart, *Naturwiss. Rundschau,* July 1964, p. 11.

[10] H. Margenau, c.s. *Physics,* McGraw-Hill, 1953, p. 599.

[11] T. Mulvey, *Discovery,* June 1964, p. 14.

[12] G. Gamow, *Biography of the Earth,* Macmillan, 1959.

[13] E. Schrödinger, *What is Life?* Cambridge, 1951.

[14] Drs. F. J. and C. M. Arnold, private communication.

[15] E. Fiala, *Bild der Wissenschaft,* Jan. 1968, p. 17.

[16] *Time,* March 11, 1966, p. 50.

[17] K. S. Davis, *Water: The Mirror of Science,* Anchor, S 18.

[18] A. Firshoff, *Discovery,* April 1965, p. 21.

[19] F. Baade, Prisma No. 435, L, Utrecht.

[20] H. C. van de Hulst, Leiden, The Netherlands (private communication).

[21] C. de Jager, *Ruimtevaart en Wetenschap,* Aula No. 352, 1968.

[22] G. T. Butler and H. Messel, Pergamon No. 174.

[23] A. B. Cambel, *Science Journal,* Oct. 1967, p. 57.

[24] E. F. J. Janetzky (private communication).

[25] M. Vertregt (private communication).

[26] K. Davis, *Scientific American,* Sept. 1965, p. 41.

[27] Basic documents are: *New Sources of Energy and Economic Development,* United Nations publication, 1957; P. Putnam, *Energy in the Future,* Van Nostrand, 1953.

[28] E. M. Wilson, *Science Journal,* July 1965, p. 50.

[29] E. Justi, *Bild der Wissenschaft,* Jan. 1967, p. 21.

[30] K. Berger, *De Ingenieur,* April 1966, p. E 45.

[31] J. R. Pierce, *Man's World of Sound,* New York, 1958. Many data are due to Prof. C. W. Koster and Ir. J. Janssen, Delft.

[32] P. Schlichting, *Kosmos,* May 1965, p. 212.

[33] G. Gamow and J. M. Clevel, *Physics: Foundation and Frontiers,* Prentice-Hall, 1960, p. 488.

[34] M. Pirenne, *Endeavour,* Oct. 1961, p. 197.

[35] J. E. Dowling, *Science,* Jan. 1967, p. 273.

[36] Several data were borrowed from R. H. Wright, *The Science of Smell,* Allen and Unwin, 1964.

[37] H. Morowitz, *New Scientist,* Feb. 1967, p. 256.

[38] R. B. Setlow and G. C. Pollard, *Molecular Biophysics*, Addison-Wesley Series, London, 1962, p. 489.
[39] J. P. van der Plaat (private communication).
[40] M. Vertregt, *Science and Religion, Past and Future*, Chapter 3 (in preparation).
[41] P. H. A. Sneath, *Discovery*, April 1964, p. 20.
[42] Partly from E. Broda, *Naturwiss. Umschau*, Aug. 1964, p. 293.
[43] Most of the data in Gullivers and Lilliputs were borrowed from E. J. Slijper, *Riesen und Zwerge im Tierreich*, Parey, Berlin, 1968.
[44] H. H. Kreutzer, *Natuur en Techniek*, Dec. 1966, p. 408.
[45] Partly from J. G. Kemeny, *Scientific American*, 1955, No. 4, p. 58.
[46] J. C. R. Licklider, *Euratom Bull.* 1967, No. 1, p. 20.
[47] G. Burbidge and F. Hoyle, *Scientific American*, Dec. 1966, p. 40.
[48] D. J. de Solla Price, *Little Science, Big Science*, Columbia Univ. Press, 1963, p. 8.
[49] J. J. Serban-Schreiber, *Le défi Americain*, Denoël, Paris 1967.
[50] P. M. S. Blackett, *Science*, 24 Feb. 1967, p. 959.
[51] Borrowed from R. Schall, Weil am Rhein.
[52] B. Chinitz, *Scientific American*, Sept. 1965, p. 134.
[53] By courtesy of A. J. M. Holmer.
[54] G. Ehrlich, *Discovery*, Nov. 1965, p. 25.
[55] G. A. M. Diepen, Delft (private communication).
[56] P. Groen, *Hemel en Dampkring*, Wolters, Gröningen, 1964, p. 199.
[57] *New Scientist*, 12 Aug. 1965, p. 401.
[58] G. Wright, *A Study of War*.
[59] *New Scientist*, 29 April 1965, p. 297.
[60] E. Neil, *New Scientist*, 3 June 1965, p. 658.
[61] H. Klensch, *Umschau in Wiss. u. Techn.*, Mai 1966, p. 310.
[62] I. Asimov, *Only a Trillion*, Allen and Unwin, London, 1962.
[63] L. B. Young, *The Mystery of Matter*, New York, 1965.
[64] *Scientific American*, Nov. 1955, p. 58.
[65] *Scientific American*, Sept. 1967, p. 102.
[66] H. Specker, *Bild der Wiss.* Oct. 1967, p. 799 and 804.
[67] *Scientific American*, Aug. 1965, p. 36.
[68] K. Egle, *Umschau in Wiss. und Technik*, 1 Sept. 1966, p. 557.
[69] *New Scientist*, 12 May 1966, p. 379.
[70] W. J. Cocke, *Phys. Review* Letters, Vol. 16, p. 662.
[71] J. Jeans, *Physics and Philosophy*, Cambridge, 1966, p. 56.
[72] Contributed by J. Jackson, Condover (England).
[73] G. Gamow, *Scientific American*, Oct. 1955.
[74] Nigel Calder, *Unless Peace Comes*, 1968.
[75] R. Giacconi, *Scientific American*, Dec. 1967, p. 6.
[76] D. M. Korevaar, Delft (private communication).

Table 1. Linear Measures

centimeters (cm)	inches	feet	yards	meters	kilometers (km)	miles
1 →	0.394	3.28×10^{-2}	1.09×10^{-2}	10^{-2}	10^{-5}	6.2×10^{-6}
2.54	← 1 →	8.33×10^{-2}	2.78×10^{-2}	2.54×10^{-2}	2.54×10^{-5}	1.58×10^{-5}
30.48	12	← 1 →	0.33	0.305	3.05×10^{-4}	1.85×10^{-4}
91.44	36	3	← 1 →	0.914	9.14×10^{-4}	5.7×10^{-4}
100	39·37	3.28	1.09	← 1 →	10^{-3}	6.21×10^{-4}
10^{5}	3.94×10^{4}	3.27×10^{3}	1.09×10^{3}	10^{3}	← 1 →	0.621
$1{\cdot}61\times10^{5}$	6.35×10^{4}	5.28×10^{3}	$1{\cdot}76\times10^{3}$	1.61×10^{3}	1.61	← 1

Table 2. Square Measures

cm^2	$inch^2$	$feet^2$	$yards^2$	m^2	acres	hectares (ha)	km^2	$miles^2$
1 →	0.155	1.1×10^{-3}	1.2×10^{-4}	10^{-4}	2.47×10^{-8}	10^{-8}	10^{-10}	3.9×10^{-11}
6.45	← 1 →	6.9×10^{-3}	7.7×10^{-4}	6.5×10^{-4}	1.6×10^{-7}	6.5×10^{-8}	6.5×10^{-10}	2.5×10^{-10}
9.3×10^{2}	144	← 1 →	0.11	9.3×10^{-2}	2.5×10^{-5}	10^{-5}	10^{-7}	3.9×10^{-8}
8.4×10^{3}	1.3×10^{3}	9	← 1 →	0.84	2.1×10^{-4}	8.5×10^{-5}	8.4×10^{-7}	3.3×10^{-7}
10^{4}	1.5×10^{3}	10.76	1.196	← 1 →	2.47×10^{-4}	10^{-4}	10^{-6}	3.9×10^{-7}
4.1×10^{7}	6.4×10^{6}	4.4×10^{4}	4.8×10^{3}	4.1×10^{3}	← 1 →	0.405	4.1×10^{-3}	1.6×10^{-3}
10^{8}	1.6×10^{7}	1.1×10^{5}	1.2×10^{4}	10^{4}	2.47	← 1 →	10^{-2}	3.9×10^{-3}
10^{10}	1.6×10^{9}	1.1×10^{7}	1.2×10^{6}	10^{6}	2.47×10^{2}	100	← 1 →	0.39
2.6×10^{10}	4×10^{9}	2.8×10^{7}	3.1×10^{6}	2.6×10^{6}	6.4×10^{2}	259	2.59	← 1

Table 3. Cubic Measures

cm^3	$inches^3$	liters (dm^3)	gallons (English)*	$feet^3$	$yards^3$	$meters^3$	(brt)	km^3	$miles^3$
← 1 →	6×10^{-2}	10^{-3}	2.2×10^{-4}	3.5×10^{-5}	1.3×10^{-6}	10^{-6}	3.5×10^{-7}	10^{-15}	2.4×10^{-16}
16.39	← 1 →	1.6×10^{-2}	3.5×10^{-3}	5.7×10^{-4}	2.1×10^{-5}	1.6×10^{-5}	5.4×10^{-6}	1.6×10^{-14}	3.8×10^{-15}
10^{3}	61	← 1 →	0.22	3.5×10^{-2}	1.3×10^{-3}	10^{-3}	3.5×10^{-4}	10^{-12}	2.4×10^{-13}
4.6×10^{3}	277	4.55	← 1 →	0.16	6×10^{-3}	4.6×10^{-3}	1.6×10^{-3}	4.6×10^{-12}	1.15×10^{-12}
2.8×10^{4}	1.7×10^{3}	28.3	6·2	← 1 →	3.7×10^{-2}	2.8×10^{-2}	10^{-2}	2.8×10^{-11}	6.8×10^{-12}
7.7×10^{5}	4.7×10^{4}	770	170	27	← 1 →	0.77	0.27	7.7×10^{-10}	1.9×10^{-10}
10^{6}	6.1×10^{4}	10^{3}	220	35·3	1.31	← 1 →	0.35	10^{-9}	2.4×10^{-10}
2.8×10^{6}	1.7×10^{5}	2.8×10^{3}	620	100	3.7	2.83	← 1 →	2.8×10^{-9}	6.9×10^{-10}
10^{15}	6.1×10^{13}	10^{12}	2.2×10^{11}	3.5×10^{10}	1.3×10^{9}	10^{9}	3.5×10^{8}	← 1 →	0.24
4.1×10^{15}	2.5×10^{14}	4.1×10^{12}	9×10^{11}	1.5×10^{11}	5.6×10^{9}	4.1×10^{9}	1.45×10^{9}	4.08	← 1

* 1 U.S. gallon = 0.83 English gallon = 3.785 liters.

Table 4. Time Units

		Seconds	Minutes	Hours
1 Hour	=	3.6×10^3	60	1
1 Day	=	86.4×10^3	1440	24
1 Year	=	31.6×10^6	52.6×10^4	8.8×10^3

Table 5. Physics

Symbol	Meaning	Value
e	Base of the natural logarithms	2.718
$\mathcal{N}$	Number of molecules in 1 grammol.	6.025×10^{23} mole^{-1}
v	Velocity	
	• of light	2.998×10^5 km sec^{-1}
	• of sound	0.340 km sec^{-1}
	• Brownian motion	2 km sec^{-1}
	• electron in orbit	2.2×10^3 km sec^{-1}
m	Rest mass	
	• electron (0.5109 MeV)	9.107×10^{-28}g
	• neutron (939.5 MeV)	1.675×10^{-24}g
	• proton (938.2 MeV)	1.672×10^{-24}g
m	Mass	
	• hydrogen atom	1.674×10^{-24}g
	• α-particle	6.64×10^{-24}g
e	charge on the electron 1.6×10^{-19} coul	4.8×10^{-10}esu
Å	Angstrom unit	10^{-8}cm
μ	micron	10^{-4}cm
mμ	millimicron	10^{-7}cm
d	Diameter	
	• H_2O molecule (in water)	2.8×10^{-8}cm
	• H atom (in H_2)	7.5×10^{-9}cm
	• electron	5.6×10^{-13}cm
	• proton	4×10^{-13}cm
g	gram	981 dynes $= 2.2 \times 10^{-3}$lb
atm	atmosphere	
	1 atm = 1033 g cm^{-2} = 10^6 dynes cm^{-2} = 760 mm Hg	
T	temperature	
	T Celsius $= 0.556\ (T_F - 32)$	
	T Fahrenheit $= 1{\cdot}8\ T_C + 32$	
	T absolute (or Kelvin) $= T_K = T_C + 273$	

Table 6. Prefixes in the Metric System

T=tera	$= 10^{12}$
G=giga	$= 10^{9}$
M=mega	$= 10^{6}$
k=kilo	$= 10^{3}$
h=hecto	$= 10^{2}$
da=deka	$= 10^{1}$
d=deci	$= 10^{-1}$
c=centi	$= 10^{-2}$
m=milli	$= 10^{-3}$
μ=micro	$= 10^{-6}$
n=nano	$= 10^{-9}$
p=pico	$= 10^{-12}$

Table 7. Space and Earth

General	
1 Light-second	$= 3\times10^{5}$ km (1.86×10^{5} miles)
1 Light-minute	$= 18\times10^{6}$ km (11.2×10^{6} miles)
1 Light-year	$= 9.5\times10^{12}$ km (5.9×10^{12} miles)
1 Parsec	= 3.3 light-years
Earth	
Distance to nearest fixed star	= 4.3 light-years
Mean distance to sun	$= 150\times10^{6}$ km (93×10^{6} miles) = 8.5 light-minutes
Mean distance to moon	$= 3.84\times10^{5}$ km (2.38×10^{5} miles) = 1.25 light-seconds
Mean diameter	$= 1.27\times10^{4}$ km (7.9×10^{3} miles)
Circumference (equator)	= 40,000 km (24,000 miles)
Surface	$= 510\times10^{6}$ km^{2} (197×10^{6} square miles)
Volume	$= 10^{12}$ km^{3} (2.4×10^{11} cubic miles)
Mass	$= 5.97\times10^{27}$ grams
Mass density	$= 5.52$ g cm^{-3}
Gravitation constant	$= 6.67\times10^{-8}$ dyne cm^{2} g^{-2}
Acceleration due to gravity (Brussels)	$= 9.81$ cm sec^{-2}
Rotation period	= 23 hours 56 min 4 sec
Sidereal period	= 365 days 5 hours 48 min 46 sec
Mean spead revolving about sun	$= 29.76$ km sec^{-1} (18.48 miles sec^{-1})
Age	= between 3 and 5×10^{9} years
Solar energy intercepted	$= 1.5\times10^{17}$ kcal hour^{-1} (1.7×10^{11} MW)
Energy consumption by mankind	$= 4.8\times10^{12}$ kcal hour^{-1} (5.5×10^{6} MW)
Water on earth	$= 1.3\times10^{9}$ km^{3}
Water in atmosphere	$= 1.3\times10^{4}$ km^{3}
Population	$= 3.5\times10^{9}$
Sun	
Diameter	$= 1.4\times10^{6}$ km (0.87×10^{6} miles)
Surface	$= 6.2\times10^{12}$ km^{2} (2.4×10^{12} square miles)
Mass	$= 33\times10^{4}\times$ mass of earth
Energy emitted	$= 3\times10^{26}$ kcal hour^{-1} $= 3.6\times10^{20}$ MW $= 3.9\times10^{33}$ erg sec^{-1}
Density	$= 1.41$ g cm^{-3}
Moon	
Diameter	$= 3.48\times10^{3}$ km (2.16×10^{3} miles)
Surface	$= 38\times10^{6}$ km^{2} (15×10^{6} square miles)
Mass	= 1/81 mass of earth
Density	$= 3.34$ g cm^{-3}

Table 8. Energy Conversion Factors

ELECTRICAL				THERMAL		MECHANICAL			PHYSICAL
eV	joules (watt-sec)	kwh	mw day	k cal	Btu	erg	kg m	ft lb	g mass equiv.
1 →	1.60×10^{-19}	4.45×10^{-26}	1.85×10^{-30}	3.83×10^{-23}	1.52×10^{-22}	1.60×10^{-12}	1.63×10^{-20}	1.18×10^{-19}	1.78×10^{-33}
6.24×10^{18}	← 1 →	2.78×10^{-7}	1.16×10^{-11}	2.39×10^{-4}	9.42×10^{-4}	10^{7}	0.102	0.738	1.11×10^{-14}
2.24×10^{25}	3.6×10^{6}	← 1 →	4.17×10^{-5}	8.60×10^{2}	3.42×10^{3}	3.6×10^{13}	3.76×10^{5}	2.66×10^{6}	4.01×10^{-8}
5.39×10^{29}	8.64×10^{10}	2.4×10^{4}	← 1 →	2.06×10^{7}	8.18×10^{7}	8.64×10^{17}	8.81×10^{9}	6.36×10^{10}	9.61×10^{-4}
2.61×10^{22}	4.19×10^{3}	1.16×10^{-3}	4.84×10^{-8}	← 1 →	3.97	4.19×10^{10}	427	3.09×10^{3}	4.66×10^{-11}
6.58×10^{21}	1.06×10^{3}	2.93×10^{-4}	1.24×10^{-8}	0.252	← 1 →	1.05×10^{10}	108	778	1.17×10^{-11}
6.24×10^{11}	10^{-7}	2.78×10^{-14}	1.16×10^{-18}	2.39×10^{-11}	9.56×10^{-11}	← 1 →	1.02×10^{-8}	7.37×10^{-8}	1.11×10^{-21}
6.12×10^{19}	9.8	2.72×10^{-6}	1.13×10^{-10}	2.34×10^{-3}	9.30×10^{-3}	9.8×10^{7}	← 1 →	7.23	1.09×10^{-13}
8.76×10^{18}	1.36	3.77×10^{-7}	1.57×10^{-11}	3.24×10^{-4}	1.29×10^{-3}	1.36×10^{7}	0.138	← 1 →	1.35×10^{-14}
5.61×10^{32}	8.99×10^{13}	2.50×10^{7}	1.04×10^{3}	2.15×10^{10}	8.51×10^{10}	8.99×10^{20}	9.17×10^{12}	6.64×10^{13}	← 1

Table 9. Power Conversion Factors

ELECTRICAL		THERMAL		MECHANICAL			
MeV sec^{-1}	kw (10^{3} joul sec^{-1})	kcal sec^{-1}	Btu sec^{-1}	erg sec^{-1}	kg m sec^{-1}	ft lb sec^{-1}	hp
1 →	1.6×10^{-16}	3.83×10^{-17}	1.54×10^{-16}	1.64×10^{6}	1.64×10^{-14}	1.18×10^{-13}	2.15×10^{-16}
6.24×10^{15}	← 1 →	0.239	0.948	10^{10}	102	738	1.34
6.61×10^{16}	4.19	← 1 →	3.97	4.18×10^{10}	427	3.08×10^{3}	5.62
6.5×10^{15}	1.05	0.25	← 1 →	1.06×10^{10}	1.08×10^{2}	778	1.42
6.24×10^{5}	10^{-10}	2.38×10^{-11}	9.47×10^{-11}	← 1 →	10^{-8}	7.35×10^{-8}	1.34×10^{-10}
6.1×10^{13}	9.8×10^{-3}	2.34×10^{-3}	9.29×10^{-3}	9.81×10^{7}	← 1 →	7.23	1.32×10^{-2}
8.46×10^{12}	1.36×10^{-3}	3.24×10^{-4}	1.28×10^{-3}	1.36×10^{7}	0.138	← 1 →	1.82×10^{-3}
4.65×10^{15}	0.745	0.178	0.707	7.45×10^{9}	76.04	550	← 1

Table 10. Technical Energy Equivalents

	kg m	kcal	kwh
1 kg standard coal	3.3×10^{6}	7000	8.5
1 kg crude oil	4.7×10^{6}	10,000	12
1 m^3 earth gas	4.2×10^{6}	9000	11
1 kg trinitrotoluene (explosive)	4.7×10^{5}	1000*	1.2
1 kg oil or fat } nutritional value	4.2×10^{6}	9000	11
1 kg protein } nutritional value	1.9×10^{6}	4100	5
1 kg carbohydrates (or sugar) } nutritional value	1.7×10^{6}	3750	4.5

* depending on the quality. Figures range between 700 and 1,100 kcal per kilogram.

Index